21世纪工业工程专业系列教材

工业工程专业英语

English for Industrial Engineering 第2版

主　编　周跃进　任秉银

本书针对工业工程专业本科生的英文学习要求，结合工业工程专业知识体系而编写，精选了工业工程专业的核心内容，题材广泛，涉及工业工程专业各领域知识。其目的是让工业工程专业学生熟悉并掌握工业工程专业的英文词汇，能够熟练阅读英文专业书籍和论文，提高英文写作能力。全书共13章，涵盖了几乎所有工业工程专业基础和专业主干课程，内容包括：工业工程概述、工作研究（包括方法研究和作业测定）、制造系统、服务系统、生产计划与控制、物流工程、人因工程、质量管理、管理信息系统、人力资源管理、科技文献检索、科技文献翻译和科技文献写作。本书内容均在英文原版教材的基础上做了改编或改写，对其中一些难句做了注释或给出了参考译文。全书收录英文专业词汇和词组500多条，书末还附有大量工业工程专业学习网站、国外著名工业工程学术及研究机构网站和参考文献。

本书可作为高等院校工业工程专业本科生的教材，也可作为工业工程专业人士的参考读物。

图书在版编目（CIP）数据

工业工程专业英语/周跃进，任秉银主编．—2版．—北京：机械工业出版社，2017.2（2023.8重印）

21世纪工业工程专业系列教材

ISBN 978-7-111-56069-2

Ⅰ.①工…　Ⅱ.①周…②任…　Ⅲ.①工业工程—英语—高等学校—教材　Ⅳ.①TB

中国版本图书馆CIP数据核字（2017）第029038号

机械工业出版社（北京市百万庄大街22号　邮政编码100037）
策划编辑：裴　泱　责任编辑：裴　泱　杨　洋
责任校对：任秀丽　封面设计：张　静
责任印制：常天培
固安县铭成印刷有限公司印刷
2023年8月第2版第6次印刷
184mm×260mm · 17.5印张 · 432千字
标准书号：ISBN 978-7-111-56069-2
定价：43.80元

电话服务　网络服务
客服电话：010-88361066　机　工　官　网：www.cmpbook.com
010-88379833　机　工　官　博：weibo.com/cmp1952
010-68326294　金　书　网：www.golden-book.com
封底无防伪标均为盗版　机工教育服务网：www.cmpedu.com

21世纪工业工程专业系列教材

编 审 委 员 会

序

每一个国家的经济发展都有自己特有的规律，而每一个国家的高等教育也都有自己独特的发展轨迹。

自从工业工程（Industrial Engineering，IE）学科于20世纪初在美国诞生以来，在世界各国得到了较快的发展。工业化强国在第一、二次世界大战中都受益于工业工程。特别是在战后经济恢复期，日本、德国等均在工业企业中大力推广工业工程的应用和培养工业工程人才，获得了良好的效果。美国著名企业家艾柯卡是美国福特和克莱斯勒汽车公司的总裁，他就毕业于美国里海大学工业工程专业。日本丰田生产方式从20世纪80年代创建以来，至今仍风靡世界各国，其创始人大野耐一的接班人——原日本丰田汽车公司生产调查部部长中山清孝说："所谓丰田生产方式，就是美国的工业工程在日本企业的应用。"韩国、新加坡、中国台湾和中国香港均于20世纪60年代起步工业工程，当时正值亚太地区经济快速发展时期。中国台湾的工业工程发展与教育是相当成功的，经过30年的努力，建立了工业工程的科研、应用和教育系统。20世纪90年代初，台湾省60所大学有48所开设了工业工程专业，至今人才需求仍兴盛不衰。更重要的是于1992年设立了工业工程学门。目前，在中国大陆的台资企业都设有工业工程部和工业工程工程师岗位，在亚太地区的学校都广泛设立工业工程专业。工业工程高水平人才的培养，对国内外经济发展和社会进步起到了重要的推动作用。

1990年6月中国机械工程学会工业工程研究会（现已更名为工业工程分会）的正式成立，以及首届全国工业工程学术会议在天津大学的胜利召开，标志着我国工业工程学科步入了一个崭新的发展阶段。人们逐渐认识到工业工程对中国管理现代化和经济现代化的重要性，并在全国范围内自发地掀起了学习、研究和推广工业工程的活动。更重要的是在1993年7月由原国家教委批准，天津大学、西安交通大学首批试办工业工程专业并招收本科生，由此开创了我国工业工程学科的先河。而后重庆大学等一批高校也先后开设了工业工程专业。时至今日，全国开设工业工程专业的院校已经有200多所，其发展速度之快，就像我国经济发展一样，令世界各国瞩目。我于2000年9月应邀赴美讲学，2001年应台湾工业工程学会邀请到台湾清华大学讲学，2003年应韩国工业工程学会邀请赴韩讲学，其题目均为"中国工业工程与高等教育发展概况"。他们均对中国大陆的工业工程学科发展给予了高度的评价，并表达了与我们保持长期交流与往来的意愿。

虽然我国工业工程高等教育自1993年就已开始，但教材建设却发展缓慢。最初，大家都使用由北京机械工程师进修学院组织编写的"自学考试"系列教材。至1998年时，全国设立工业工程专业的高校已达三四十所，但仍没有一套适用的专业教材。在这种情况下，工业工程分会与中国科学技术出版社合作出版了一套工业工程专业教材，并请西安交通大学汪应洛教授任编委会主任。这套教材的出版有效地缓解了当时工业工程专业高等教育教材短缺的压力，对我国工业工程专业高等教育的发展起到了重要的推动作用。

然而，近年来我国工业工程学科发展十分迅猛，开设工业工程专业的高校数量直线上升，同时教育部也不断出台新的政策，对工业工程的学科建设、办学思想、办学水平等进行规范和评估。在

新的形势下，为了适应教学改革的要求，满足全国普通高等院校工业工程专业教学的需要，机械工业出版社推出的这套“21 世纪工业工程专业系列教材”是十分及时和必要的。在教材编写启动会上，编审委员会组织国内工业工程专家、学者对本套教材的学术定位、编写思想、特色进行了深入研讨，力求在确保高学术水平的基础上，适应普通高等院校教学的需求，做到适应面广、针对性强、专业内容丰富。同时，本套教材还将配备课件，相应的实验、实习教程，案例教程以及企业现场录像，实现立体化。尽管如此，由于工业工程在我国正处于快速成长期，加上我们的学术水平和知识有限，教材中难免存在各种不足，恳请国内外同仁批评指正。

教育部工业工程类专业教学指导委员会主任
中国机械工程学会工业工程分会主任

于天津

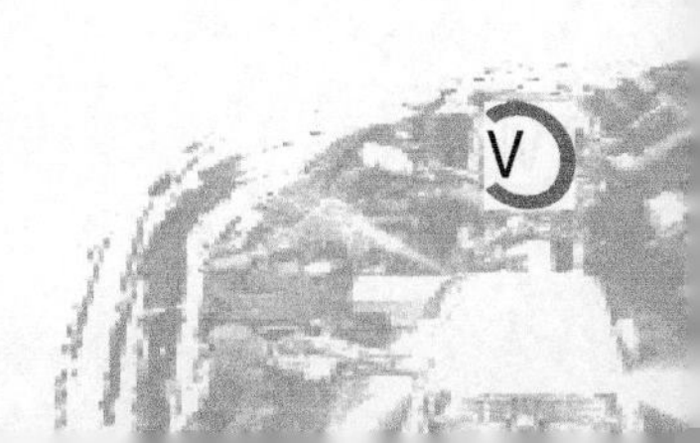

前　　言

本书是为工业工程专业学生学习专业英语而编写的。 其目的是加强工业工程专业学生的英语训练， 使其能掌握工业工程专业的英语词汇， 顺利阅读工业工程专业英语文献， 提高与国外同行的学术交流和交往水平。

本书共 13 章， 涵盖了几乎所有工业工程专业基础和专业主干课程， 内容包括： 工业工程概述、 工作研究 （含方法研究和作业测定）、 制造系统、 服务系统、 生产计划与控制、 物流工程、 人因工程、 质量管理、 管理信息系统、 人力资源管理、 科技文献检索、 科技文献翻译和科技文献写作。 此外， 还附有工业工程领域著名的学术机构及协会组织的名称、 网站以及工业工程方面的常见词汇等。

本书内容丰富， 在教学安排上各学校 （各任课教师） 可根据学生的英语水平和学校对该课程的课时要求灵活安排， 其中有些内容可作为学生课外阅读材料。

本书的特色是： ① 针对性强。 本书针对工业工程专业学生介绍与该专业基础课程和专业课程有关的英语基础知识和专业知识。 ② 方便学生阅读。 本书附有大量注释和词汇以帮助学生自己阅读。 ③ 包含思考题和习题。 它们方便读者练习和复习， 考查其掌握本书基本理论和方法的程度。

本书第 1 章和第 2 章第 2、 3、 4 单元由周跃进编写， 第 2 章第 1 单元和第 5 章由马力编写， 第 3 章和第 4 章由李萍编写， 第 6 章和第 13 章由赵丽娟编写， 第 7 章、 第 8 章和第 10 章由任秉银编写， 第 11 章由李迁编写， 第 9 章和第 12 章由李建辉编写。 全书由周跃进统稿， 唐任仲教授主审。

在本书的编写过程中， 陈卉、 翁元、 刘聆哲、 时楠、 王佳、 胡冬伟、 王天博等研究生做了大量工作， 在此表示衷心感谢！

本书涉及工业工程领域相当广泛的内容， 由于编写时间较紧， 加之作者英语水平所限， 书中难免出现疏漏或不当之处， 敬请广大读者和同仁提出宝贵意见， 以便今后再版时加以改进。

周跃进
于南京大学南园

Contents

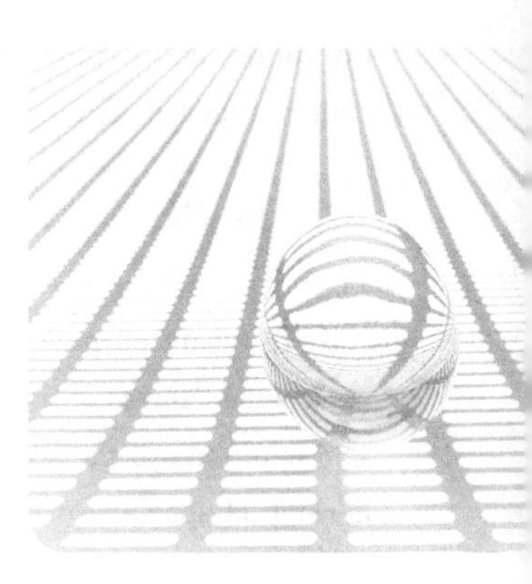

Chapter 1
Introduction to Industrial Engineering

Unit 1 What is Industrial Engineering?

The Roles of IE

Industrial engineering (IE) is emerging as one of the classic and novel professions that will be counted for solving complex and systematic problems in the highly technological world of today.[1] In particular, with the transformation and promotion of China's economy and China's reform and opening deep-going, the demand for IE will increase and widen continuously and urgently.

A production system or service system includes inputs, transformation, and outputs. Through transformation, the added values are increased and the system efficiency and effectiveness are improved. Transformation processes rely on the technologies used and management sciences as well as their combination.

Managing a production system or service system is a challenging and complex task—one that requires the knowledge of fundamental sciences, engineering sciences, behavioral sciences, computer and information sciences, economics, and a great number of topics concerning the basic principles and techniques of production and service systems.

The Demand for IE Graduates

Industrial engineering curricula are designed to prepare the students to meet the challenges of the future for the construction of Chinese economic and harmonious society. Many IE graduates (IEs) will, indeed, design and run modern manufacturing systems and facilities. Others will select to engage in service activities such as health-care delivery, finance, logistics, transportation, education, public administration, or consulting and so on.

The demand for IEs is strong and growing each year. In fact, the demand for IEs greatly exceeds the supply. This demand/supply imbalance is greater for IE than for any other engineering or science disciplines and is projected to exist for many years in the future. Therefore, over 250 universities or colleges opened IE program in China in 2015.

The Objectives of the Textbook

The main purpose of this textbook is to introduce systematical theories and advanced techniques and methodologies of the relevant subjects of industrial engineering as well as their English expression. The other aim of the textbook is to strengthen and improve student's the ability of reading and comprehension of specialized English literatures related to industrial engineering.

Engineering and Science

How did the two words "industrial" and "engineering" get combined to form the term "industrial engineering"? What is the relationship between industrial engineering and other engineering disciplines, to business management, or the social sciences?

To understand the role of industrial engineering in today's economic and knowledge-based era, it is beneficial to learn the historical developments that were hopeful in the evolution of IE. There are many ways to write a historical development of engineering. The treatment in this unit is brief because our interest is in reviewing the significances of engineering development, particularly those leading to industrial engineering as a specialty. More complete histories are available in the reference [2]—[4].

Engineering and science have developed in a parallel, complementary fashion, although they are not always at the same pace. Whereas science is concerned with the quest for basic knowledge, engineering is concerned with the application of scientific knowledge to the solution of problems and to the quest for a "better life". Obviously, knowledge cannot be applied until it is discovered, and once discovered, it will soon be put to use. In its efforts to solve problems, engineering provides feedback to science in areas where new knowledge is needed. Thus, science and engineering work hand in hand.

Engineering Applications—Tools

Although "science" and "engineering" each have distinguishing features and are regarded as different disciplines, in some cases a "scientist" and an "engineer" might be the same person. This was especially true in earlier times when there were very few means of communicating basic knowledge. The person who discovered the knowledge also put it to use.

We naturally think of such outstanding accomplishments as the pyramids in Egypt, the Great Wall of China, the Roman construction projects, and so on, when we recall early engineering accomplishments. Each of these involved an impressive application of fundamental knowledge.

Just as fundamental, however, were accomplishments that are not as well known. The inclined plane, the bow, the wheel, the corkscrew, the waterwheel, the sail, the simple lever, the many, many other developments were very hopeful in the engineer' s efforts to provide a better life.

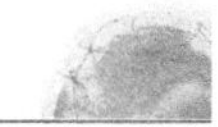

Engineering Basis

Almost all engineering developments prior to 1800 had to do with physical phenomena: such as overcoming friction, lifting, storing, hauling, constructing, and fastening. Later developments were concerned with chemical and molecular phenomena: such as electricity, properties of materials, thermal processes, combustion, and other chemical processes.

Fundamental to almost all engineering developments were the advances made in mathematics. Procedures for accurately measuring distances, angles, weights, and time were necessary for almost all early engineering accomplishments. As these procedures were refined, greater accomplishments were realized.

Another very important contribution of mathematics was the ability to represent reality in abstract terms. A mathematical model of a complex system can be manipulated such that relationships between variables in the system can be understood. The simple relationship commonly called the Pythagorean theorem is such an example. This theorem says that the hypothenuse of a right triangle is equaled as the square root of the sum of the squares of the adjacent sides. The use of abstract models representing complex physical systems is a fundamental tool of engineers.

As a final comment on early development, let us take care of an early development that did not come. The missing early development is related to the behavioral sciences. The understanding of human behavior has lagged greatly behind developments in the mathematical, physical, and chemical sciences. This is important to industrial engineers because the systems designed by IEs involve people as one of the basic components. The lack of progress in behavioral science has impeded the industrial engineer in his efforts to design optimal systems involving people.

The Modern Era of Engineering

Based on the book *Introduction to Industrial and Systems Engineering*, it defines the *modern era* of engineering as beginning in 1750, even though there were many important developments between 1400 and 1750. There are two reasons to choose 1750 as the beginning of modern engineering:

(1) Engineering schools appeared in France in the eighteenth century.

(2) The term *civil engineer* was first used in 1750.

Civil Engineering

Principles of early engineering were first learned in military colleges and were concerned primarily with road, bridge and fortifications construction. This kind of academic training was referred to as military engineering. When some of the same principles were applied to nonmilitary attempts, it was natural to call them as civilian engineering, or simply civil engineering.

Mechanical Engineering

With the development of civil engineering, the relevant disciplines had also developed. Interrelated advancements in the fields of physics and mathematics set up the groundwork for practical applications of mechanical principles. A significant advancement was the development of a practical steam engine that could accomplish useful work. Once such an engine was available (approximately 1700), many mechanical devices were developed that could be driven by the engine. These efforts culminated in the emergence of mechanical engineering as a distinct branch in the early nineteenth century.

Electrical Engineering

The discovery and applications of electricity and magnetism are another example of such advancement, which were the fundamental work done in the later part of the eighteenth century. Although early scientists had known about magnetism and static electricity, an understanding of these phenomena did not start until Benjamin Franklin's famous kite-flying experiment in 1752. In the next 50 years the foundation of electrical science was built up primarily by German and French scientists.

The first distinguishing application of electrical science was the development of the telegraph by Samuel Morse. Morse telegraph is a kind of telegraph that sends messages using dots and dashes or short and long sounds or flashes of light to represent letters of the alphabet and numbers. Thomas Edison's invention of the carbon-filament lamp (which is still used today) led to widespread use of electricity for lighting purposes. This, in turn, spurred very rapid developments in the generation, transmission, and utilization of electrical energy for a variety of labor-saving purposes. Engineers who chose to specialize in this field were naturally called electrical engineers.

Chemical Engineering

Along with the developments in mechanical and electrical technologies were accompanying developments in the understanding of substances and their properties. The science of chemistry came up, which is concerned with understanding the nature of matter and in learning how to produce desirable changes in materials. Fuels were required for the new internal combustion engines being developed. Lubricants were needed for the rapidly growing array of mechanical devices. Protective coatings were required for houses, metal products, ships, and so forth. Dyes were needed in the manufacture of a wide variety of consumer products. Somewhat later, artificial materials were required to carry out certain functions that could not be performed as well or at all by natural materials. This field of engineering effort naturally became known as chemical engineering.

Industry

After making clear of science and engineering, it is time to talk about the term "industry". A clear indication of the way in which human effort has been harnessed as a force for the commercial production of goods and services is the change in meaning of the word industry. Coming from the Latin word *industria*, meaning "diligent activity directed to some purpose" and its descendant, old French

industrie, with the senses "activity" "ability" and "a trade or occupation", our word (first recorded in 1475) originally meant "skill" "a device" and "diligence" as well as "a trade". As more and more human effort over the course of the Industrial Revolution became involved in producing goods and services for sale, the last sense of industry as well as the slightly newer sense "systematic work or habitual employment" grew in importance, to a large extent taking over the word. We can even speak now of the Shakespeare industry, rather like the garment industry. The sense "diligence, assiduity", lives on, however, perhaps even to survive industry itself. From the origin of this word, we can find that it means a variety of economic or social activities.

Large Scale Production

As industrial organizations emerged to make use of the rapidly developing array of technological innovations, the size and complexity of manufacturing units increased dramatically. Large scale production was made possible through three important concepts:

(1) Interchangeability of parts.

(2) Specialization of labor.

(3) Standardization.

Through large scale production the unit cost of consumer products was reduced dramatically and productivity was increased substantially.

The Origination of IE

The base was now laid for a dramatic shift in the lifestyles and cultures of industrialized countries. Within nearly twenty years the People's Republic of China and other developing countries changed largely rural, agricultural economies and societies to urban, industrialized economies and societies. The suddenness of this change is probably the cause of many of today's urgent problems, for example, air pollution, enviromental change, resources shortage, diseases increase, traffic crowding and population aging, and so on.

During the early part of this movement it was recognized that business and management practices that had worked well for small shops and farms simply were inadequate for large, complex manufacturing organization. The need for better management systems led to the development of what is now called-industrial engineering.

The Definition of Industrial Engineering

The following formal definition of industrial engineering has been adopted by the Institute of Industrial Engineers (IIE):

Industrial engineering is concerned with the design, improvement, and installation of integrated systems of people, materials, information, equipment, and energy. It draws upon specialized knowledge and skill in the mathematical, physical, and social sciences together with the principles and methods of engineering analysis and design to specify, predict, and evaluate, the results to be obtained from such systems.[2]

As used in this context, the term *industrial* is intended to be interpreted in the most general way as mentioned above. Although the term *industrial* is often associated with manufacturing organization, here it is intended to apply to any organization. The basic principles of industrial engineering are being applied widely in agriculture, education, hospitals, banks, government organizations, and so on.

Read this definition again. Imagine any large factory that you have seen in which thousands of workers, hundreds of machines, a large variety of materials and thousands and millions of yuans must be combined in the most productive, cost-effective manner. Think about a large city that also requires millions of workers, millions of vehicles and other machinery, materials, and thousands and millions of yuans in order to deliver services required by public. Imagine how much more effectively the city could be run if the principles of industrial engineering were applied.

Notes

1. Industrial engineering (IE) is emerging as one of the classic and novel professions that will be counted for solving complex and systematic problems in the highly technological world of today.

句意：作为一种古老而又新颖的专业，工业工程的出现将用来解决当今高度技术发展的世界所遇到的复杂的系统问题。

classic 意味着工业工程是一个古老的专业，有100多年的历史；novel 意味着工业工程一直致力于改革和创新，紧跟时代步伐并致力于解决当今社会现实问题。

2. Industrial engineering is concerned with the design, improvement, and installation of integrated systems of people, materials, information, equipment, and energy. It draws upon specialized knowledge and skill in the mathematical, physical, and social sciences together with the principles and methods of engineering analysis and design to specify, predict, and evaluate, the results to be obtained from such systems.

句意：工业工程是对由人员、物料、信息、设备和能源所组成的集成系统进行设计、改善和设置的一门学科。它综合运用数学、物理学和社会科学方面的专门知识和技术以及工程分析和设计的原理与方法，对该系统所取得的成果进行鉴定、预测和评价。

Exercises

1. What is science? What is engineering? What is industry? What is industrial engineering?
2. Simply describe the formulation of industrial engineering.
3. Explain the definition of industrial engineering.
4. Explain the origination of industrial engineering.
5. What are the focus topics of the five big disciplines?
6. What is the base of engineering and why?

Unit 2 History of Industrial Engineering

The Formulation of Industrial Engineering

Industrial engineering emerged as a profession as a result of the industrial revolution and the accompanying need for technically trained people who could plan, organize, and direct the operations of large complex systems. The need to increase efficiency and effectiveness of operations was also an original stimulus for the emergence of industrial engineering. Some early developments are explored, in order to understand the general setting in which industrial engineering was born. For more details, see the excellent work of Emerson and Naehring—*Origins of Industrial Engineering: the Early Years of a Profession*.

Now let us recall some famous people who contributed their efforts for IE.

Charles Babbage's Division of Labor

Charles Babbage (1792 – 1871)[1] visited factories in England and the United States in the early 1800's and began a systematical record of the details involved in many factory operations. For example, he observed that the manufacture of straight pins involved seven distinct operations. He carefully measured the cost of performing each operation as well as the time per operation required to manufacture a pound of pins. Babbage presented this information in a table, and thus demonstrated that money could be saved by using women and children to perform the lower-skilled operations. The higher-skilled, higher-paid men need only perform those operations requiring the higher skill levels. Babbage published a book containing his findings, entitled *On the Economy of Machinery and Manufactures* (1832). In addition to Babbage's concept of division of labor, the book contained new ideas on organizing and very advanced (at that) concepts of harmonious labor relations. Significantly, Babbage restricted his work to that of observing and did not attempt to improve the work methods or to reduce the operation times.

Eli Whitney's Interchangeable Concept

The concept of interchangeable manufacture was a key development leading to the modern system of mass production. This concept was to produce parts so accurately that a specific part of a particular unit of a product could be interchanged with the same part from another unit of the product, with no degradation of performance in either unit of the product. Eli Whitney (1765 – 1825)[2] received a government contract to manufacture muskets using this method. His another contribution was the design and construction of new machines that could be operated by laborers with a minimal amount of training. Through the successful application of these two concepts, Whitney created the first mass production system.

In the period of around 1880 industrial operations were conducted in a much different manner from today. There was very little planning and organizing, as such. A first line supervisor was given verbal instructions on the work to be done and a crew of (usually) poorly trained workers. The supervisor was expected to his men make as hard as he could. Any improved efficiency in work methods usually came from the worker himself in his effort to find an easier way to get his work done. There was virtually no attention given to overall coordination of a factory or process.

Frederick Winslow Taylor's Efficiency Improvement

Frederick Winslow Taylor (1856 – 1915)[3] is credited by recognizing the potential improvements to be gained from analyzing the work content of a job and designing the job for maximum efficiency. Taylor's original contribution, constituting the beginning of industrial engineering, was his three-phase method of improving efficiency: Analyze and improve the method of performing work, reduce the times required, and set standards for what the times should be. Taylor's method brought about significant and rapid increases in productivity. Later developments stemming from Taylor's work led to improvement in the overall planning and scheduling of an entire production process.

Frank B. Gilbreth's Motion Study

Frank B. Gilbreth (1868 – 1924) extended Taylor's work considerably. Gilbreth's primary contribution was the identification, analysis, and measurement of fundamental motions involved in performing work. By classifying motions as "reach" "grasp" "transport" and so on, and by using motion pictures of workers performing their tasks, Gilbreth was able to measure the average time to perform each basic motion under varying conditions. This permitted, for the first time, jobs to be designed and the time required to perform the job known before the fact. This was a fundamental step in the development of industrial engineering as profession based on "science" rather than "art".

Lillian Gilbreth's Human Relations Study

Dr. Lillian Gilbreth (1878 – 1972), wife of Frank, is credited by bringing to the engineering profession a concern for human welfare and human relations. Having received a doctoral degree in psychology from Brown University, Dr. Gilbreth became a full partner with her husband in developing the foundational concepts of industrial engineering. During Dr. Gilbreth's long life, she witnessed and contributed to the birth, growth, and maturation of the IE profession. She became known as the "first lady of engineering" and the "first ambassador of management". She received many, many honors and awards from professional organizations, universities, and governments around the world. She was the first woman to be elected to the National Academy of Engineering[4].

Gantt Chart

Another early pioneer in industrial engineering was Henry L. Gantt (1861 – 1919), who devised the so-called Gantt chart. The Gantt chart was a significant contribution in which it provided a systematic graphical procedure for preplanning and scheduling work activities, reviewing progress, and updating the schedule. Gantt charts are still in widespread use today.

Shewhart's Statistical Quality Control

Walter A. Shewhart (1891 – 1961) developed the fundamental principles of statistical quality control in 1924. This was another important development in providing a scientific base to industrial engineering practice.

Many other industrial engineering pioneers contributed to the early development of the profession. During the 1920s and 1930s much fundamental work was done on economic aspects of managerial decisions, inventory problems, incentive plans, factory layout problems, material handling problems, and principles of organization. Although these pioneers are too numerous to be mentioned in this brief chronology, more complete historical accounts are available elsewhere[6-9].

Scientific Management and IE

The significance of the early development of industrial engineering is far greater than most people realize, and in fact, ranks among the greatest achievements of all time. In 1968, Peter Drucker described the importance of "scientific management", which we today call industrial engineering:

Scientific management (we today probably call it "systematic work study", and eliminate thereby many misunderstandings the term has caused) has proved to be the most effective idea of the last century. It is the only basic American idea that has had worldwide acceptance and influence. Whenever it has been applied, it has raised the productivity and with it the earnings of the manual worker, and especially of the laborer, while greatly reducing his physical efforts and his hours of work. It has probably multiplied the laborer's productivity by a factor of one hundred.

Evolution of Industrial Engineering

Shown in Figure 1-1 are a number of significant events and developments that have occurred in the evolution of industrial engineering. The position of each event relative to the time axis is intended only to show the approximate time at which that event occurred. In many cases, such as time studies, the event on the chart merely indicates its beginning; time studies are still a fundamental tool of industrial engineering.

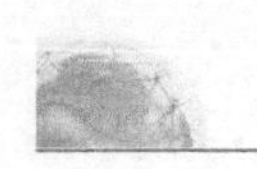

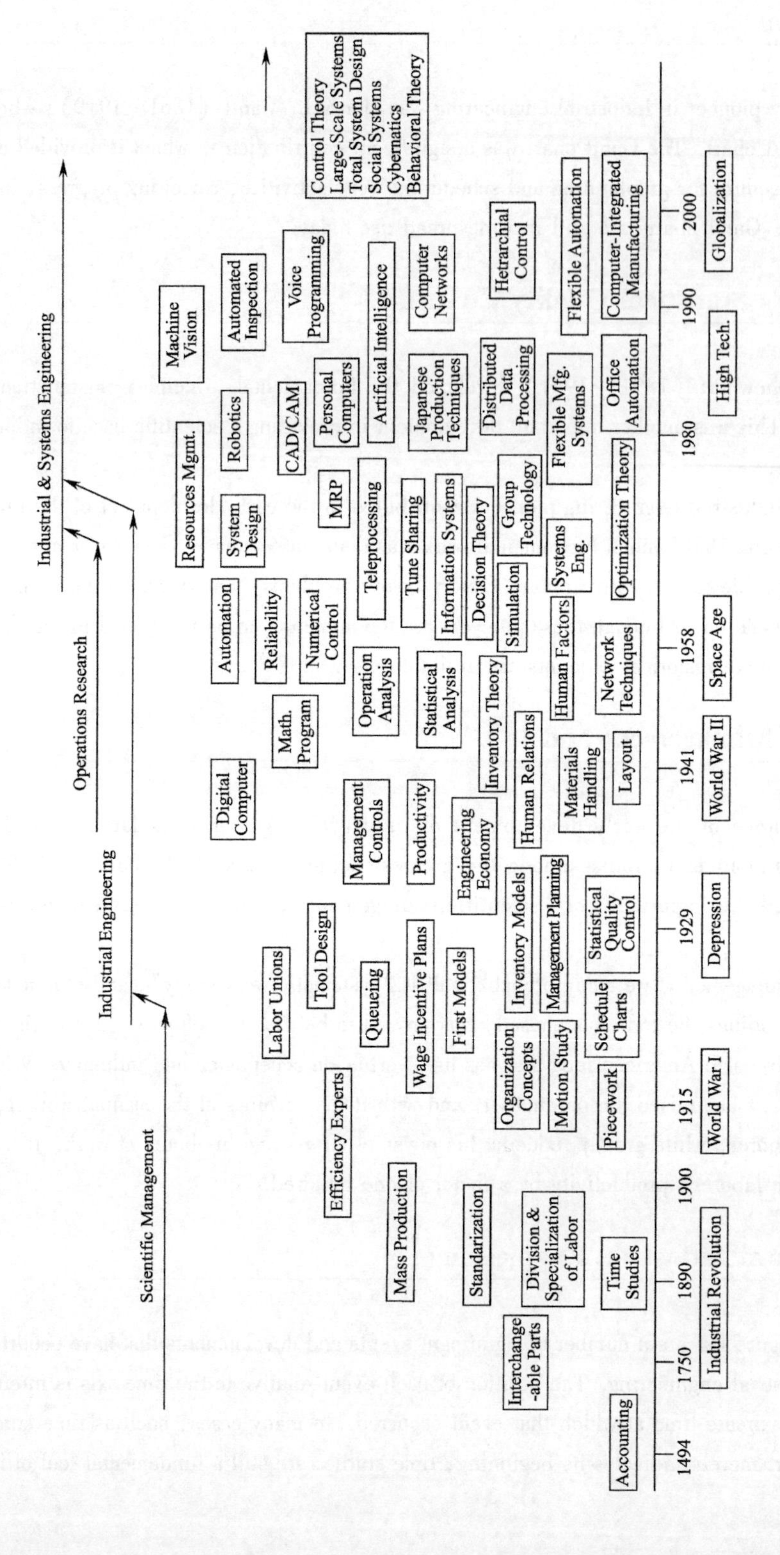

Figure 1-1　Chronology of Significant Events and Developments in the Evolution of Industrial Engineering

（Source：Reference[1]）

At the top of the chart are shown four overlapping periods of emphasis. The labels are somewhat arbitrary. The period from approximate 1900 through the mid-1930s is generally referred to as scientific management. The next period, labeled industrial engineering, begins in the late 1920s and is shown extending to the present time. The period marked operations research had a great influence on industrial engineering practice and is shown beginning in the mid-1940s and extending somewhat past the mid-1970s. The fourth period, industrial and systems engineering, is shown beginning around 1970 and extending indefinitely into the future.

Notice that each period mixes into the succeeding period. In this context, a particular period never really "ends". Instead, the cumulative periods simply result in the attainment of the particular characteristics of following periods. The influences of Taylor and other pioneers certainly continue to the present time and are reflected in industrial engineering practices.

The authors believe that the IE profession has evolved to the point that a significant new direction has been emerging for the last decade. We imagine this new orientation as not only building on the previous foundations of industrial engineering and operations research, but also including appropriate concepts from feedback control theory, computer and information science, behavioral theory, financial engineering, systems engineering, and cybernetics, and so on.

Some of the most exciting and productive years for our profession are just ahead, specially, it is true for Chinese IE students. The opportunities for service to humankind have never been greater.

Characteristics of Industrial Engineering

Although there can never be general agreement on this point, the vocations that are commonly referred to as professions are medicine, teaching, architecture, law, ministry, and engineering. Smith points out that these vocations have four common characteristics:

(1) Associated with a profession is a significant body of special knowledge. [5]

(2) Preparation for a profession includes an internship-like training period following the formal education. [6]

(3) The standards of a profession, including a code of ethics, are maintained at a high level through a self-policing system of controls over those practicing the profession.

(4) Each member of a profession recognizes his responsibilities to society over and above responsibilities to his client or to other members of the profession. [7]

Other definitions of a profession go further and state that its members must engage in continued study and that its prime purpose must be rendering of a public services. Although these matters are subject to interpretation, they are evidence that professionalism is indeed a serious matter and that much is expected of people who claim to be professionals.

Engineer's Tasks

Engineers are frequently involved in decisions that have a profound effect on society. The design of particular devices almost always involves the safety of the user. The design of the processes frequently affects the environment. The design and location of a factory affect the community and its citizens. The design of a management system great affects the individuals working for the organization—their comfort, their sense of worth, their financial status, and so on.

Engineers have been unjustly accused of creating almost all the current problems in our society, for example, pollution and crowding. The truth is that engineers have worked diligently and persistently to provide society with the things the social scientists claimed were needed. Pollution, crowding, and unsightly freeways are simply the byproducts.

Engineers do, however, have a tremendous responsibility to protect the public welfare. They must continually engage in trade-offs between costs and factors affecting public welfare, such as safety. An automobile can be made almost completely safe for its occupants, but the automobile would cost approximately ¥1,000,000 and would weigh approximately 14 tons. Even if someone were willing to pay this much for an automobile, there are other considerations: the safety of people in lighter cars traveling on the same roadways and the fuel consumption of a 14-ton vehicle.

Engineers are also frequently caught in a controversy between their employer or client and the public. Companies must produce items for sale at competitive prices, even though a "cheap" design may mean an unsafe or unreliable product.

The engineering profession enjoys a very favorable reputation regarding its adherence to professional ethics. Fortunately, early engineers were perceptive and conscious of the public good. The National Society of Professional Engineers[8] publishes a set of Canons of Ethics for engineers. They do not answer every question or resolve every controversy, but they do provide a good foundation from which one may extrapolate to cover almost any situation.

Functions of Industrial Engineering

Traditionally, IE functions have been concentrated at the operations level of a firm. Other firms, recognizing the broad-based skills of their IEs have expanded their activities to include the design of management systems. In recent years, as the IE role has added a systems flavor, IEs are expected to engage in activities at the corporate level. To gain an appreciation from the broad spectrum of activities in which an IE might be engaged, we present in the Table 1-1 formed a list activities grouped according to the three categories: production operations, management systems, and corporate services.

Table 1-1 The Functions of IE

Production operations	Management systems	Corporate services
Related to the product or service	Related to information systems	Relative to comprehensive planning
Related to the process of manufacturing the product or producing the service	Related to financial and cost systems	Relative to policies and procedures
Related to the facilities	Related to personnel	Relative to performance measurement
Related to work methods and standards		Relative to analysis
Related to production planning and control		

Notes

1. Charles Babbage, 1792 – 1871, British mathematician and inventor of an analytical machine based on principles similar to those used in modern digital computers.

2. Eli Whitney, 1765 – 1825, American inventor and manufacturer whose invention of the cotton gin revolutionized the cotton industry. He also established the first factory to assemble muskets with interchangeable parts, marking the advent of modern mass production.

3. Frederick Winslow Taylor, 1856 – 1915, American inventor, engineer, and efficiency expert noted for his innovations in industrial engineering and management.

4. National Academy of Engineering 美国工程科学研究院

5. Associated with a profession is a significant body of special knowledge.

句意：与一个专业相关的是该专业所具有的特殊的知识结构。

6. Preparation for a profession includes an internship-like training period following the formal education.

句意：准备从事一个专业包括在正规教育阶段之后有一个类似训练的实习期。

7. Each member of a profession recognizes his responsibilities to society over and above responsibilities to his client or to other members of the profession.

句意：专业中的每一个成员都能认识到他对社会的责任是在他对客户或其他专业成员的责任之上的。

8. The National Society of Professional Engineers 美国国家专业工程师协会

Exercises

1. Explain the characteristics of industrial engineering as a profession and professional ethics.

2. Indicate the functions of industrial engineering by using practical cases.

3. Describe how the work of Taylor was related to that of Gilbreth, and vice versa.

4. Find case studies that illustrate the applicability of industrial engineering principles and methodologies to non-manufacturing systems, such as banks, hospitals, and government functions.

5. Develop a detailed chronology of the development of fundamental IE concepts during the period 1880 – 1930 using the references and other materials that you may find in your library or website.

6. Describe the tasks of engineers.

Unit 3 Academic Disciplines of Industrial Engineering

Five Big Major Engineering Disciplines and Their Development

In United States, there are five big major engineering disciplines (civil, chemical, electrical, industrial, and mechanical) that were the branches of engineering that emerged prior to the time of World War I. These developments were part of the Industrial Revolution that was occurring worldwide, and the beginning of the technological revolution is still occurring.

Development following World War Ⅱ led to other engineering disciplines, such as nuclear engineering, electronic engineering, aeronautical engineering, and even computer engineering. The Space Age led to astronautical engineering. Recent concerns with the environment have led to environmental engineering and bioengineering. These newer engineering disciplines are often considered specialties within one or more of the "big five" disciplines of civil, chemical, electrical, industrial, and mechanical engineering.

It is the same situation of United States that industrial engineering belongs to the first-class disciplines in China since 2013. However, it may belong to different colleges, such as engineering college, management college, or other college.

The Beginning of IE Subjects

Topics that later evolved into industrial engineering subjects were initially taught as special courses in mechanical engineering departments. The first separate departments of industrial engineering were established at Pennsylvania State University and at Syracuse University in 1908. (The program at Syracuse was short-lived, but it was reestablished in 1925.) An IE option in mechanical engineering was established at Purdue University in 1911. A rather complete history of industrial engineering academic programs may be found in *Origins of Industrial Engineering: the Early Years of a Profession.*

The practice of having an IE option within mechanical engineering departments was the predominant pattern until the end of World War Ⅱ, and separate IE departments were established in colleges and universities throughout the last century.

There was very little graduate level work in industrial engineering prior to World War Ⅱ. Once separate departments were established, master's and doctor's level programs began to appear.

Education of Modern IE—Sub-disciplines

Today, more than over, industrial engineering means different things to different people. In fact, one of the ways to develop an understanding of modern industrial engineering is by gaining an understanding of both its sub-disciplines and how it relates to other fields. It would be convenient, for purposes of explanation, if there were clearly defined boundaries between sub-disciplines and related fields to industrial engineering; unfortunately, that is not the case. The fields most commonly referred to today as sub-disciplines of or related to industrial engineering are management science, economics, statistics, operations research, ergonomics, manufacturing engineering, systems engineering logistic engineering and computer and information science. There are those in each of these disciplines who believe their field is separate and distinct from industrial engineering.

The education of the modern industrial engineers involves some combination of content from all the disciplines just mentioned. In any particular instance, the combination depends on the industrial engineering academic department and the company in which individuals gain work experience. What may or may not be apparent at this point is the diversity of courses offerings in industrial engineering.

Whereas depth in a single discipline is the primary strength of an electrical, mechanical, or civil engineering degree program, breadth of understanding across a broad range of related topic areas, both within and outside the college of engineering, as well as depth in industrial engineering subjects, is the primary strength of an industrial engineering degree program.

The following introduction to each of these sub- and related disciplines is intended to offer both relevant history and a limited comparative understanding of the present nature of each discipline.

Management Science

Of all the disciplines mentioned above, management science, simply called management, was one of the earliest to emerge in human history. If management is the art and science of directing human effort, then it must have begun when one person attempted to get another person to work. There is considerably less than unanimity of opinion today as to how best to do that.

Recognition of the need for planning, organizing, leading, rendering, and controlling human effort can be traced back at least as far as early Egyptian times. The execution of these functions is essential if, for example, one is to build a pyramid in a reasonable amount time.

With the possible exception of an introductory statement or paragraph about pre-twentieth-century management thought, most modern texts in management being their development with a discussion of the scientific concepts of Taylor. Many authors refer to Taylor as the "father of scientific management", whereas others call him the "father of industrial engineering".

There is little question that the subdivision of management commonly referred to as production management has a great deal in common with industrial engineering. In most business colleges, production management is a sequence of one to two courses at the undergraduate level that attempt to familiarize management students with concepts and techniques specific to the analysis and management of production activity. Industrial engineering, on the other hand, is an engineering degree curriculum concerned with analysis, design, and control of productive systems. A productive system is any system that produces either a product or a service. Production management courses are often primarily concerned with teaching management students how to manage (i. e., direct human efforts) in a production environment, with less attention paid to the analysis and design of productive systems.

Industrial engineering students, on the other hand, are taught primarily how to analyze and design productive systems and the control procedures for efficiently operating such systems. Except for a possible course or two concerned with fundamental understanding of management concepts for directing the human effort associated with such systems, it is generally assumed that industrial engineers will not operate the systems they design. The training of a race car driver is analogous to management education: the design of the car is analogous to industrial engineering education. The race car driver wants to know first and foremost how to drive the car and its less concerned with a detailed understanding of how it works. The industrial engineer designs the car with a driver in mind but with no intention of getting behind the wheel on the day of the race. [1] The engineer does intend to be there, however, to observe the performance of the car and assist with appropriate adjustments. The engineer's concern after the initial design is with design improvement or the continued development of procedures that result in optimum performance.

In industrial engineering program, the courses involved management include:

- Management science;
- Production management;
- Logistic management/Supply chain management;
- Management information systems;
- Human resources management;
- Project management;
- Quality management.

Economics

Economics is the study of how societies choose to use scarce productive resources that have alternative uses, to produce commodities of various kinds, and to distribute them among different groups. The essence of economics is to acknowledge the reality of scarcity and then figure out how to organize society in a way which produces the most efficient use of resources. That is where economics makes its unique contribution.

Economists, therefore, study how people make decisions: how much they work, what they buy, how much they save, and how they invest their savings. Economists also study how people interact with one another. Finally, economists analyze forces and trends that affect the economy as a whole, including the growth in average income, the fraction of the population that cannot find work, and the rate at which prices are rising.

Economists use the scientific approach to understand economic life. This involves observing economic affairs and drawing upon statistics and the historical record. For complex phenomena like the impacts of budget deficits or the causes of inflation, historical research has provided a rich mine of insights. Often, economics relies upon analyses and theories. Theoretical approaches allow economists to make broad generalizations, such as those concerning the advantages of international trade and specialization or the disadvantagesof tariffs and quotas. In addition, economists have developed a specialized technique known as econometrics, which applies the tools of statistics to economic problems. Using econometrics, economists can sift through mountains of data to extract simple relationships.

The field of economics is traditionally divided into two broad subfields: microeconomics and macroeconomics. Microeconomics is the study of how households and firms make decisions and how they interact in specific markets. It focuses on the behavior of the units—the firms, households, and individuals—that make up the economy. It is concerned with how the individual units make decisions and what affects those decisions.

Macroeconomics is the study of economy wide phenomena. It looks at the behavior of the economy as a whole, in particular the behavior of such aggregate measures as the overall rates of unemployment, inflation, and economic growth and the balance of trade. The aggregate numbers do not tell us what any one firm or household is doing. They tell us what is happening in total, or on average.

A microeconomist might study the effects of rent control on housing in New York City, the impact

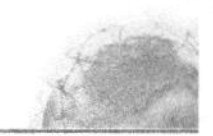

of foreign competition on the U. S. auto industry, or the effects of compulsory school attendance on workers' earnings. A macroeconomist might study the effects of borrowing by the federal government, the changes over time in the economy's rate of unemployment, or alternative policies to promote growth in national living standards.

Microeconomics and macroeconomics are closely intertwined. Because changes in the overall economy arise from the decisions of millions of individuals, it is impossible to understand macroeconomic developments without considering the associated microeconomic decisions.

Statistics

Statistics has been and will continue to be distinct from industrial engineering. However, the approach of industrial engineering has changed significantly; the world around us is viewed as probabilistic in nature rather thandeterministic. By deterministic it is meant that all actions under consideration in a particular study situation are assumed to be certain. Probabilistic implies that at least one aspect of the study situation has a probability of occurrence associated with it that must be considered.

In a deterministic problem you may assume, for example, that the cost of a used car is ¥20, 000. All calculations concerning buying the car would assume the fixed ¥20, 000 cost. In a similar probabilistic problem, you may assume that there is an 80 percent chance that the car can be purchased for ¥20, 000, and a 20 percent chance that it can be bought for ¥15, 000.

The probabilistic view of the world has so pervaded industrial engineering practice and education that a beginning course in probability and statistics has now become the most important prerequisite in a typical industrial engineering degree program. Industrial engineering has been leading the way for other engineering disciplines in this development, and it seems likely that the improved insight it offers to problems will ultimately result in all disciplines shifting toward a more probabilistic view of the world.

Operations Research (O. R.)

Operations research has been defined by the Operational Research Society of the United Kingdom as follows:

The attack of modern science on complex problems arising in the direction and management of large systems of men, machines, materials, and money in industry, business, government, and defense. The distinctive approach is to develop a scientific model of the system, incorporating measurement of factors such as chance and risk, with which to predict and compare the outcomes of alterative decisions, strategies, or controls. The purpose is to help management determine its policies and actions scientifically.[2]

The definition says that O. R. is applicable just about anywhere there are systems that need to be managed. A "scientific model" is perhaps the key phrase in the definition. This implies that unless a scientific model (usually mathematical) is developed, it cannot be called operations research. The next phrase says that the objectives are to "predict and compare the outcomes of alterative decisions, strategies, or controls". This implies that any scientific model that predicts or evaluates the results of decisions or policies is operations research. Finally, the overall objective of operations research is to "help management determine its policies and actions scientifically". It is significant that we say to help management. No O. R. tool "makes" decisions. It is only an aid to the decision maker.

Let us review the definition of IE in the first unit. It is obvious that IE and O. R. have commonalities. O. R. and IE indeed do have many of the same objectives and work on many of the same problems. The primary difference is that O. R. has a higher level of theoretical and mathematical orientation, providing a major portion of the science base of IE.

O. R. carries a connotation of mathematical orientation, but IE does not restrict itself to any specific approach. Figure 1-2 illustrates the relationship between the two disciplines in terms of mathematical sophistication.

Many industrial engineers work in the area of operations research, as do mathematicians, statisticians, physicists, sociologists, and others. It is significant to observe that many O. R. programs in universities are taught by industrial engineering faculty, or by mathematics and statistics faculty, or by Operations Research faculty.

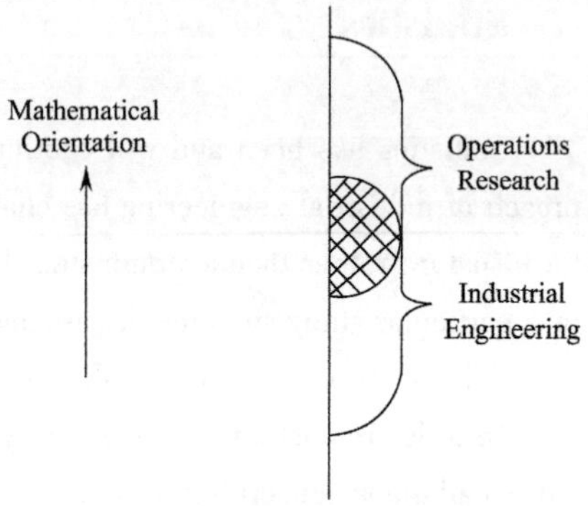

Mathematical Orientation

Operations Research

Industrial Engineering

Figure 1-2 Industrial Engineering and Operations Research

Nature of O. R. is in mathematical involvement. Research into new mathematical techniques could be called O. R. but not IE, although the research is often done by industrial engineers.

Let's explore the nature of O. R. in an attempt to categorize the techniques. In every situation that we wish to model using an O. R. approach we encounter the problem of estimating the values of the parameters (factors such as the price of raw material or time to produce a part). Those parameters may not be constant over time. That is, they may behave as random variables or perhaps change in some predictable fashion. One approach is to forget that they are random variables and use a mathematical model that does not recognize variation. This approach is called the deterministic approach and is often used. Actually, all models we have seen thus far have been deterministic. If the model recognizes this random variation, the approach is called probabilistic approach.

Ergonomics

Ergonomics previously called *human factors*, is one of the subdisciplines of industrial engineering, closely associated with both industrial and experiment psychology. The field of psychology has produced a wealth of information and theory about the human body and mind that is readily available to human factors engineers. Industrial engineering systems by nature are often human-machine systems, in contrast to hardware systems in electrical engineering, for example. The design of human-machine systems involves determining the best combination of human and machine elements. A typical course summaries the considerable research that has been performed to date in national ergonomics laboratories (i. e. , Wright Patterson Air Force Base) and in universities (e. g. , Texas Technological University,

Virginia Polytechnic Institute and Tsinghua University of China). These ergonomics accomplishments assist in familiarizing the industrial engineering student with human-machine systems design. The text Human Factors Engineering by McCormick has been used extensively for this purpose. A significant amount of ergonomics research is now being performed in industrial engineering departments, complementing the continuing research that has been underway for many years in industrial psychology.

Ergonomics studies physiological aspects and psychological aspects of human beings. Physiological aspects of human beings include anthropometry and impacts of working environment. Psychological aspects of human beings include industrial psychology, mental health, and incentive mechanism. More detailed contents see Chapter 7 in this book.

Systems Engineering

1. System Concept

We repeat a word "system" during our discussion. What is a system? A system may be defined as a set of components which are related by some form of interaction, and which act together to achieve some objective or purpose. In this definition, components are simply the individual parts, or elements, that collectively make up a system. Relationships are the cause-effect dependencies between components. The objective or purpose of a system is the desired state or outcome which the system is attempting to achieve.

Systems may be classified in a number of different ways. We discuss a few classifications that illustrate the similarities and dissimilarities of systems.

Natural vs. Man-Made Systems—Natural systems are those that exist as a result of processes occurring in the natural world. A river is an example of a natural system. Man-made systems are those that own their origin to human activity. A bridge built to cross a river is an example of a man-made system.

Static vs. Dynamic Systems—A static system is one that has structure but no associated activity. The bridge crossing a river is a static system. A dynamic system is one that involves time varying behavior. The Chinese economy is an example of a dynamic system.

Physical vs. Abstract Systems—A physical system is one that involves physically existing components. A factory is an example of a physical system, because it involves machines, buildings, people, and so on. Abstract systems are those in which symbols represent the system components. An architect's drawing of a factory is an abstract system, consisting of lines, shading, and dimensioning.

Open vs. Closed Systems—An open system is one that interacts with its environment, allowing materials (matter), information, and energy to cross its boundaries. A closed system operates with very little interchange with its environment.

Industrial engineers design systems at two levels. The first level is calledhuman activity systems and is concerned with the physical workplace at which human activity occurs. The second level is called management control systems and is concerned with procedures for planning, measuring, and controlling all activities within the organization.

2. Cybernetics

Two highly significant works were published in 1948; one was Norbert Wiener's[3] *Cybernetics*, or *Control and Communication in the Animal and the Machine*, and the other was Claude Shannon's *The Mathematical Theory of Communication*. Wiener derived the word *cybernetics* from a Greek word meaning steersman, and his subject was the generality of negative feedback in systems spanning the biological and physical world.

The most commonly used example of negative feedback is the thermostat. When the temperature drops sufficiently below some desired value, the thermostat initiates the heating portion of the cycle, and the heat is added until a temperature is reached that is greater than the desired temperature. Heating is then stopped to permit cooling to negate the overheating. Negative feedback means that some action is taken to oppose or negate an unacceptable difference.

Figure 1-3 is a conceptual model of negative feedback in management systems. An apparent condition is compared with a goal, and if asufficient difference (i. e., error) exists, management action is taken to reduce the difference. The action should result in a change in the apparent condition so that later comparison of the apparent condition with the goal will cause the controlling action to cease. Assume that a manufacturer wishes to have 100 units of inventory on hand. After reviewing his inventory status he notes that he has only 80. If 20 is a sufficient difference, he would probably perform the management action of ordering more material in an attempt to raise his inventory level closer to the desired level. The order, after a purchase delay, should bring the apparent condition closer the goal, removing the need for additional management action. This concept is consistent with the exception principle of managerial control, which says that management attention should be directed to situations in which abnormal values are known to exist. It is management's job to cause an undesired value to return to a normal or steady-state level.

It is the generality of this concept and other characteristics of systems that make Wiener's text significant. Homeostasis is a word commonly employed in the biological sciences in connection with the regulatory processes in living organisms. Analogous regulation can be identified in such diverse systems as water flow in irrigation and current flow in electrical networks.

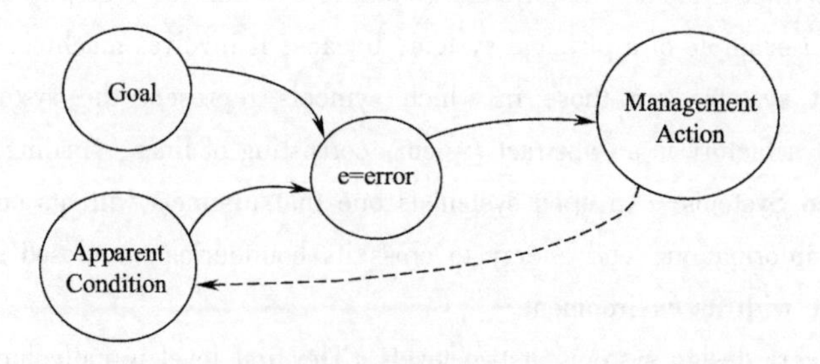

Figure 1-3　Negative Feedback in Management Systems

Source: Reference

3. A General Systems Theory

Wiener's work is generally considered to be the starting point of what is now commonly referred to as general systems theory. Schlager reported in 1956 on the basis of a review of a nationwide survey, that the first known use of the term systems engineering was in the Bell Telephone Laboratories in the early 1940s. Considering the problems the Bell System faced at that time in expanding its system, it is understandable that the term might well have been received there. The RCA (Radio Corporation of America) Corporation, in the years just preceding this, had recognized the need for a systems engineering point of view in the development of a television broadcasting system.

In 1946 the newly created RAND Corporation developed a methodology that they labeled systems analysis. Quade and Boucher in *Systems Analysis and Policy Planning* defined systems analysis as "a systematic approach to helping a decision-maker choose a course of action by investigating his full problem, searching out objectives and alternatives, and comparing them in the light of their consequences, using an appropriate framework—in so far as possible analytic—to bring expert judgment and intuition to bear on the problem".

4. The Differences between Systems Engineering and Operations Research

Some fairly clear differences have emerged over the years between operations research and systems engineering. Although the early philosophers of operations research believed it to be the beginning of an analytical attack, via mathematics, on large-scale problems, a review of the operations research literature shows that for most problems, the number and complexity of representations must be limited if analytically sound solutions are to be reached. Some operations research problems involve a large number of equations—some linear programming solutions, for example—but the complexities of representation in any one of the many equations may, and often do, make the entire set of equations unsolvable. For many problems today the techniques of operations research offer solutions that were unavailable in the recent past.

Systems engineering seems to have developed with less dependence on "hard" mathematical representation of all aspects of a system. Digital simulation is a much more frequently employed technique in systems engineering, particularly if the system cannot be tightly represented and solved analytically because there is no appropriate analytical technique or the data are not in the form required for a specific operations research technique.

Systems demand that a macro perspective be attained in effectively dealing with any significant problem. There is a considerable danger to attempting to solve a problem without first getting the big picture of the total system in which the problem is embedded. You may mess up the system in the process of fixing the problem—it is commonly called "winning the battle, but losing the war".

Manufacturing Engineering

Manufacturing Engineering may be defined as designing the production process for a product.

The Society of Manufacturing Engineers has represented manufacturing engineers in the United States since 1932. Manufacturing engineering is a familiar industrial function name in manufacturing

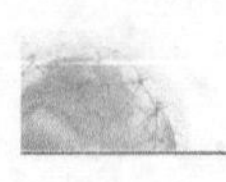

organizations but has never been as well established as an academic degree program in U. S. universities as industrial engineering, for example. There are more than 70 industrial engineering and 10 manufacturing engineering academic programs in the United States. However, it is very popular and almost every engineering university or college has manufacturing engineering academic program in China.

Industrial engineering and manufacturing engineering are distinct and typical complementary functions in a manufacturing organization. Most firms need both functions represented in their organizations to be truly effective. If one tries to substitute one function for the other, the function omitted typically represents a weakness in that manufacturing organization that will likely limit the overall capacity of the technical effort in that organization.

A typical manufacturing engineering department is composed of numerous technical professionals (mechanical engineers, thermodynamicists, material engineers, computer scientists, etc.). Each professional represents some part of the technical processes in use at that manufacturing plant. For example, the thermodynamicists may concern herself with product fin design for dissipating heat, an electrical engineer may worry about test sets and related procedures, and the chemical engineer may concern himself with solution concentrations and related specifications for plating processes. The processes function, as they were intended to function because the manufacturing engineering department represents the assemblage of technical expertise necessary to keep all the manufacturing processes under control.

If that is what the manufacturing engineering department does, why do we need an industrial engineering department as well? The core of a typical industrial engineering department is a more homogeneous collection of professionals, typically make up of industrial engineers, with and without degrees, and technicians/technologists. It may well include other specialists, however, with degrees or experience in psychology, management, computer science, and statistics, as well as other engineering disciplines. The smallest entity that an industrial engineer typically deals with is a machine. The machine to an industrial engineer is a black box that has a production rate, yield rate, required operator skills, process capabilities, and other production system attributes. The industrial engineer is concerned with developing a production system that produces the required quantity of products at an appropriate cost and quality. If a machine does not work properly, he may refer the problem to maintenance, and if they cannot fix it, he may refer it to the manufacturing engineering department. If they cannot fix it, they should design a machine that will work for the required step in the production system under design. More detailed contents see Chapter 3 in this textbook.

Logistics Engineering

Logistics engineering is a field of engineering dedicated to the scientific organization of the purchase, transport, storage, distribution, and warehousing of materials and finished goods.

Logistics is that part of the supply chain process that plans, implements, and controls the efficient, effective flow and storage of goods, services, and related information from the point of origin to the point of consumption in order to meet customers' requirements.

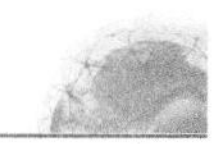

Logistics engineering involves planning and analysis of logistic systems, facilities location and layout design, logistic equipment, design of material handling systems, planning and design of warehouses, transport, and distribution. More detailed contents see Chapter 6 in this textbook.

Computer and Information Science

It is Information Age—a time when knowledge is power. Today, more than ever, businesses are using information technology to gain and sustain a competitive advantage. Information technology (IT) adding industrial engineering (IE) becomes contemporary industrial engineering (CIE), which is effective tool for the reform and business of firms. The related courses include the foundation of computers, data structure, the principles and applications of database, and management information systems (MIS).

Whether your major is what, you are preparing to enter the business world as a knowledge worker.

In the Information Age, MIS are vitally important tools and topics.

MIS deal with the planning for, development, management, and use of information technology tools to help people perform all tasks related to information processing and management.

IT is any computer-based tool that people use to work with information and support the information and information processing needs of an organization.

Information is a key resource in an organization. Data come from equipment, production, inventory, procurement, sales, product design, and after service. The data through processing become useful information which supports decision making. Finally the information integrated forms the business intelligence and helps make decisions automatically.

Industrial engineering particularly takes IT as its major tools in the process of enterprise management and improvement from strategic, tactical, to operational levels. More detailed contents see Chapter 9 in this textbook.

Notes:

1. The industrial engineer designs the car with a driver in mind but with no intention of getting behind the wheel on the day of the race.

句意：工业工程师设计赛车时关心坐在车里的赛车手如何驾驶，而不关心比赛时赛车如何运动。

2. The attack of modern science on complex problems arising in the direction and management of large systems of men, machines, materials, and money in industry, business, government, and defense. The distinctive approach is to develop a scientific model of the system, incorporating measurement of factors such as chance and risk, with which to predict and compare the outcomes of alterative decisions, strategies, or controls. The purpose is to help management determine its policies and actions scientifically.

句意：运筹学是用来处理产业、商业、政府和国防领域由人、机器、物料和资金组成的复杂的大系统的管理和指导的现代科学。运筹学所采用的独特方法是建立科学的系统模型，

再加上机会与风险等的因素评价，对所选择的决策、战略和控制的结果进行预测和比较。运筹学的目的是帮助管理层决定其政策并科学地采取行动。

3. Norbert Wiener, 1894 - 1964, American mathematician who founded the field of cybernetics.

Exercises

1. What is manufacturing? What is manufacturing engineering?

2. Why do we emphasize on manufacturing engineering as a fundamental of industrial engineering?

3. Explain the similarities and dissimilarities between operations research and systems engineering.

4. Explain the similarities and dissimilarities between operations research and industrial engineering.

5. Describe the sub-disciplines of industrial engineering and involved research contexts.

6. Indicate the dissimilarities between other engineering disciplines and industrial engineering and the dissimilarities between business management and industrial engineering.

Unit 4 Development of Industrial Engineering

Industrial Engineering Responsibility

The industrial engineering responsibility involves the integration of workers, machines, materials, information, capital, and managerial know-how into a producing system that will produce the right product, at the right cost, at the right time. Manufacturing engineering technical talent is one of the underlying technical plant-supporting resources that guarantee success of that production system.

In summary, it is necessary to know the technical details of each of the processes (i. e., manufacturing engineering) and then integrate all the elements of a producing system (workers, materials, equipment, information, etc.) so that a quality product is made at the right time and cost (i. e., industrial engineering).

What, then, are modern industrial engineers? First and foremost, they are engineers. They must take the following courses (Table 1-2). There are some differences depending on the different universities. Industrial engineering students, more than any other engineering students, are in classes in other departments in other colleges in various remote corners of a campus. Industrial engineers have a far broader training than students in other engineering disciplines. That training is probably their greatest asset when it comes time to leave campus, as most students must sooner or later, and go to work.

Table 1-2 Involved Disciplines of Industrial Engineering

Disciplines	Courses	Skills
Mathematics	Advanced Mathematics Engineering Mathematics Numerical Calculation	Technical Interpersonal Consulting
Mechanical Engineering	Engineering Graphics and Computer Graphics Fundamentals of Machine Design Engineering Mechanics Material Mechanics	
Physics	College Physics Experiments of College Physics	
Electrical Engineering	Electrical Engineering and Electronic Techniques Fundamental of Control Theory and Techniques	
Computer Science	Fundamentals of Computer Technology C + + Program Design Data Structure Principles and Applications of Microcomputer Principles and Applications of Database	
Information Science	Management Information Systems Electronic Commerce Enterprise Resources Planning	
Manufacturing Engineering	Manufacturing Systems Contemporary Manufacturing Systems Production Automation Technology of Numerical Control Product Design and Rapid Prototyping Manufacturing	
Management	Management Science Production Management Supply Chain Management Quality Management Management of Enterprise Strategy Project Management Human Resources Management Organizational Behavioral Science Industrial Psychology	
Statistics	Management Statistics	
Systems Engineering	Introduction of Systems Engineering Systems Simulation	
Finance & Economics	Economics Finance andAccounting Engineering Economics Cost Management Finance Engineering International Finance and Trading	
Operations Research	Operations Research Advanced Operations Research	
Ergonomics	Human Factors Engineering	
IE Professional	Fundamentals of Industrial Engineering Logistics Engineering IE Seminars	

In China, the formal IE education began in 1993. Unlike the United States at that time, it is a secondary subject under the discipline of management science and engineering. It can be granted engineering degrees or management degrees. So this program is opened in both engineering schools (or other related schools) and management schools. However, IE became the first-class subject in 2013 in China. The significant symbol is the establishment of IE teaching steering committee at that year.

The industrial engineering profession evolved from other engineering disciplines. Figure 1-2 illustrates the general nature of this evolution as well as the inputs to industrial engineering from certain nonengineering disciplines.

Industrial engineering organizations are shown in Table 1-3.

Table 1-3 Industrial Engineering Organizations

Organizational Name	Established Date	Remark
The Taylor Society	1915	
The Society of Industrial Engineers (SIE)	1917	
The Society for the Advancement of Management	1936	Combined with the first two
The American Institute of Industrial Engineers (AIIE)	1948	The Journal of IE IE/IIE, Transactions
The Institute of Industrial Engineers (IIE)	1981	Including 80 countries
The Chinese Institute of Industrial Engineering	1990	The Instiente of Industrial and Systems Engineers
American Society of Mechanical Engineers (ASME)	1880	
The American Management Association (AMA)	1922	
The Society of Manufacturing Engineers (SME)	1932	
The American Manufacturers Association	1929	
Operations Research Society of American	1938	
American Society for Quality Control (ASQ)	1946	
American Production and Inventory Control Society (APICS)	1957	
Society of American Value Engineers (SAVE)	1947	
American Association of Cost Engineers (AACE)	1956	
Robotics Society of America (RSA)	1972	

Challenges for Industrial Engineering

The remarkable achievements of the engineering profession have, ironically, contributed to some of the problems now facing human society. In the petroleum industry, engineers have designed highly efficient processes and systems for locating, extracting, processing, and distributing petroleum products. [1] Increased efficiency led to increased usage. This process could continue indefinitely if there were unlimited amounts of oil to be found. The oil shortages of the latest years have made the world painfully aware that petroleum reserves are indeed limited.

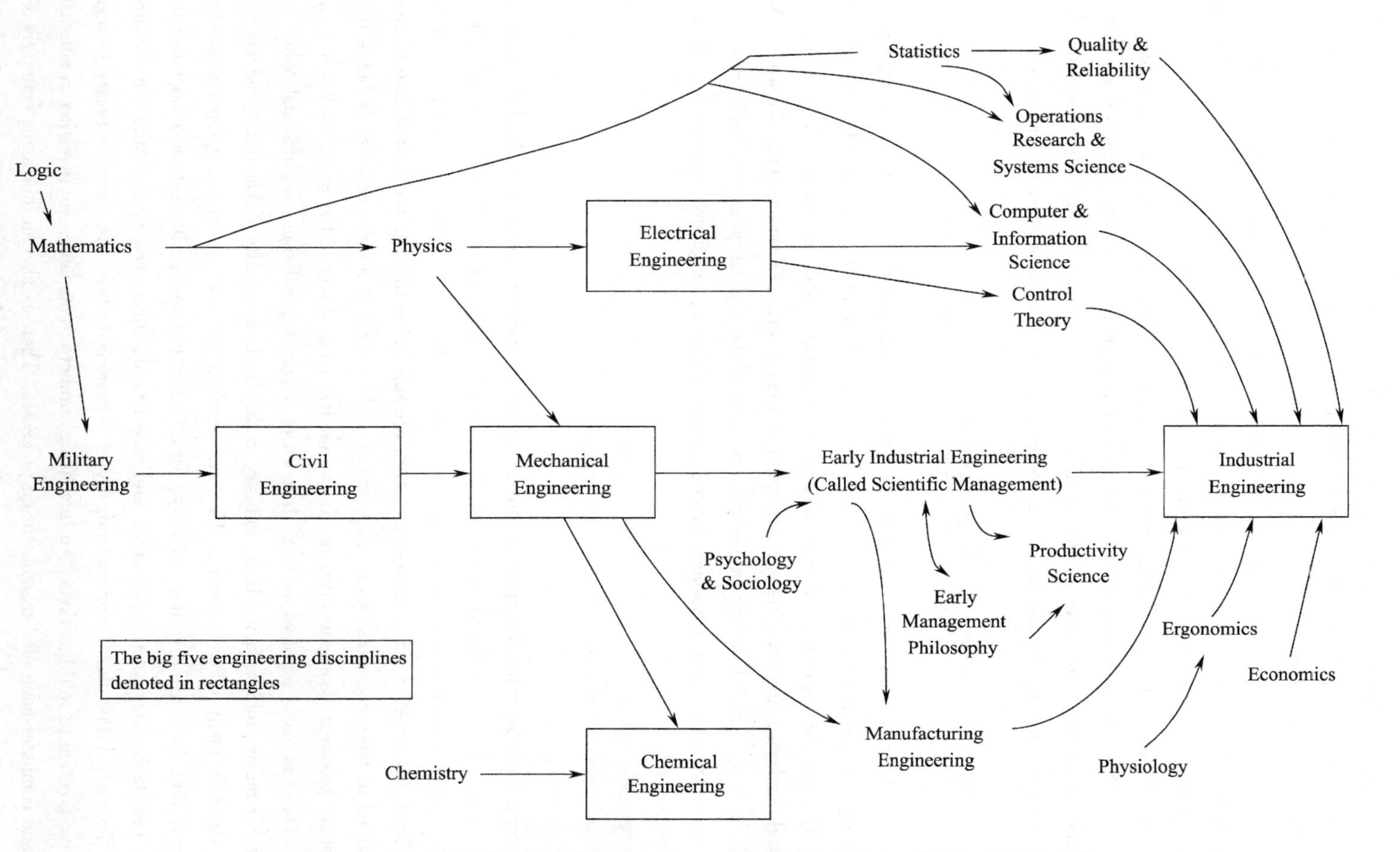

Figure 1-4 Relationship of Industrial Engineering to Other Engineering and Scientific Disciplines

Similar examples could be cited for other natural resources. The world has a finite amount of non-renewable resources. One of the major challenges for future engineers is to learn to accomplish the engineer's mission in recognition of these constraints and how to save the existing resources during production and services activities.

Another major challenge facing the engineering profession is to design systems and processes that are compatible with our natural environment. Dumping wastes into a river or allowing harmful gases to escape into the atmosphere are no longer permissible design strategies.

A major challenge that future engineers will encounter is that of designing products that are safe and reliable. More and more companies are being held accountable for faulty products that result in injury or harm to the buyer or to the public. Some places are now enacting legislation requiring that the design of all products affecting the public be approved bya registered professional engineer[2].

The Efforts of Industrial Engineering

The facts have been proved that through the application of sound industrial engineering practices, American industry became the strongest industry in the world during the two decades following World War Ⅱ. Through the application of technological advancements and the never-ending search for improved work methods, American industry attained the highest level of productivity in the world. American products were regarded in the world market as being of the highest quality and reliability. The industrial engineering practices of Europe, Japan, Taiwan of China, Singapore, and the Republic of Korea also proved it is true.

The Problems Facing U. S. Industries

Productivity is the ratio of outputs to inputs. American industry continues to lead the world in labor productivity, but other countries are rapidly closing the gap. Japan, China, Germany, France, and other European countries are outpacing the United States in the annual rate of productivity increase. Since labor cost in many of these countries are still below those in the United States, products manufactured in these countries can be sold on the world market at prices considerably below those of United States products. Consequently, the U. S. has lost large shares of the world market in many industries. This has had a negative on U. S. balance of payments, inflation, and national debt.

Another major problem facing U. S. industry in the declining quality of American products relative to those made in many other countries. This has resulted, in part, from the aging equipment and processes in their factories. American industry, for a variety of reasons, has not made adequate investments in new technology, new equipment, and new manufacturing concepts. There are encouraging signs in the early 1990s that American industry is beginning to take the steps necessary to regain a competitive position in world markets. An increasing number of companies are investing in automation, robots, and computer-controlled manufacturing processes. There is also an increasing emphasis on improved quality. The critical importance of a well-educated work force is also being recognized.

There is a problem which is on the revitalization planning of manufacturing in United States. Since

1980s U. S. manufacturing industries all went outside because of the limitations of land, worker force, and resources. The revitalization planning of manufacturing hopes to return the position of a giant country in manufacturing. It also proves how important it is to manufacture for a country!

The Problems and Opportunities Facing China

Compared with the United States, China is still a developing country. The more problems are faced by Chinese industry at present including the problems that faced by U. S. industry in the past and other problems such as the adjustment of industrial structures, the reform of supply sides, the restructuring of state-owned enterprises, the increase of product quality and productivity, the establishment and improvement of market order, and so on.

The business magazines are now saying that an "industrial revolution and informationalization" is under way in Chinese industry. The engineering profession, particularly industrial engineering, will have a major role to play in this movement.

As management grapple with the many problems facing China, a realization will soon emerge that their solutions require an input from technically competent professionals. Perhaps the greatest challenge facing the engineering profession is to become involved in political issues and to provide assistance to lawmakers in the design of social systems.

Despite the many remarkable achievements of engineers being made in the past, their greatest contributions to society will occur in the future. They will be expected to design systems (manufacturing and service systems) that optimally utilize available resources for the satisfaction of human needs. Thus, the basic mission of the industrial engineer remains essentially the same as always. Additional constraints, however, such as resources limitations, environmental concerns, and political reform issues, must be considered as an integral part of the engineering process.

Notes

1. In the petroleum industry, engineers have designed highly efficient processes and systems for locating, extracting, processing, and distributing petroleum products.

句意：在石油工业，工程师们已经设计出高度有效的工艺和系统来勘探、提炼、加工和分销石油产品。

2. a registered professional engineer 一个注册专业工程师

Exercises

1. What about the development of IE in China and other countries?
2. What are the problems facing United States at present?
3. What are the problems facing China at present?
4. What are the opportunities and challenges for Chinese industrial engineering?
5. Why do you want to select industrial engineering as your profession?
6. What are the responsibilities of industrial engineers?

Chapter 2 Work Study

Introduction to Work Study

Industrial engineering covers techniques aimed at improving productivity. And work study is the important part of the classical industrial engineering, which regards the work or operations systems that are the microcosmic foundations of the production system as the research objects. It is the earliest technique of industrial engineering, its basic function is to diagnose and analyze the existing production system, and the final purpose is to improve productivity. The work study is an integrative technology of engineering and management which can increase the ratio of output without investment or with little investment, so it has been paid the general attention of the industrial circle for a long time. [1]

Work study includes two parts: method study and work measurement. Figure 2-1 illustrates the basic research contents of work study.

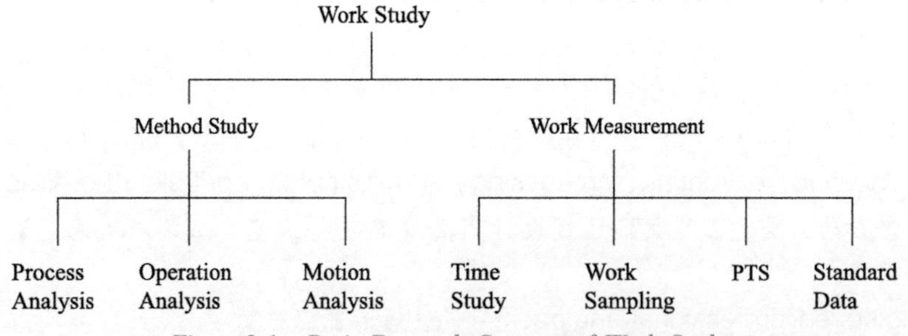

Figure 2-1 Basic Research Contents of Work Study

Unit 1 Method Study

The Definition and Content of Method Study

In this unit we will discuss the method study mainly. We can define method study as the systemat-

ic recording and critical examination of ways of doing things in order to make improvements. From the Figure 2-1, we can see that the main contents of the method study include process analysis, operation analysis and motion analysis.

Steps of Method Study

Selecting—the work to be studied & boundaries to be defined. Method study focuses on certain key operational types of work that should be targeted: such as bottlenecks, poor use of materials, poor layout, inconsistencies on quality, highly fatiguing tasks, high levels of unexplained employee complaints.

Recording—the relevant facts of the job to be recorded. The main recording techniques are by charts: including operation process chart, flow process chart, flow diagram, man-machine process chart, gang process chart and two-hand process chart.

Examining and developing—the way the job to be performed. In this stage researchers should analyze the original system and design new methods to suit needs and abilities of operators, and these methods should be the most practical, economic, and effective for that job.

Six questioning techniques (5W1H). Six questioning techniques are the primary tools, by which the critical examination is conducted. In the Table 2-1, we list the details of the six questioning techniques.

Analysis with "ECRS" principles. When we design new methods for the current work or system, we should use the following principles flexibly:

- Eliminate: When we could not find the satisfied answers through questioning "What has been done?" "Why is this operation necessary?" "Why is this operation performed in this manner?" these operations belong to unnecessary parts so they should be eliminated.
- Combine: We should try to combine these operations or motions that could not be cancelled.
- Rearrange: By the elimination and combination, we can rearrange the work according the three questions "who, where and when" to get the greatest order and eliminate the repetitions.
- Simplify: The last idea is simplification. The simplest method and facilities are used to save the manpower, time and money after the above three necessary steps.

Table 2-1 Six Questioning Techniques

	Primary Questions		Secondary Questions	
Purpose	What is actually done?	Why is the activity necessary at all?	What else might be done?	What should be done?
Place	Where is it being done?	Why is it done at that particular place?	Where else might it be done?	Where should it be done?
Sequence	When is it done?	Why is it done at that particular time?	When might it be done?	When should it be done?
Person	Who is doing it?	Why is it done by that particular person?	Who else might do it?	Who should do it?

(Contained)

	Primary Questions		Secondary Questions	
Means	How is it being done?	Why is it done that particular way?	How else might it be done?	How should it be done?

Setup—Setting up new standard work methods.

Maintaining—New method is maintained until it becomes the accepted norm.

The Classifications of Method Study

There are three main classes of method study: process analysis, operation analysis and motion analysis.

Process Analysis

Process analysis is one of the main contents of method study, by which researchers can completely observe and record the whole production processes and carry out integrated analysis from a macroscopic viewpoint. According to the different subjects, process analysis can be divided into four kinds: operation process analysis, flow process analysis, layout and the path analysis, and management transaction analysis; all kinds of analysis tools used for correlation analysis chart, as shown in Table 2-2.

Table 2-2 Process analysis types and the tools

Process analysis types	Process analysis tools
Operation process analysis	Operation process chart
Flow process analysis	Flow process chart
Layout and the path analysis	Flow diagramand string diagram
Management transaction analysis	Management transaction flow diagram

Operation Process Chart

An operation process chart is a process chart giving the general flow of all components in a product with symbols of operations and inspections. The chart in itself is an ideal plant layout since each step is shown in its proper chronological sequence. It can record the sequence of operations and inspections, activity times and description of each activity and indicate the main activities involved in tasks. Using this chart people can gain understanding of tasks, eliminate unnecessary tasks or combine tasks. The convention of operation process chart is illustrated in the Figure 2-2.

Flow Process Chart

There are five symbols that present flow process chart, as shown in Table 2-3. They are also called ASME symbols.

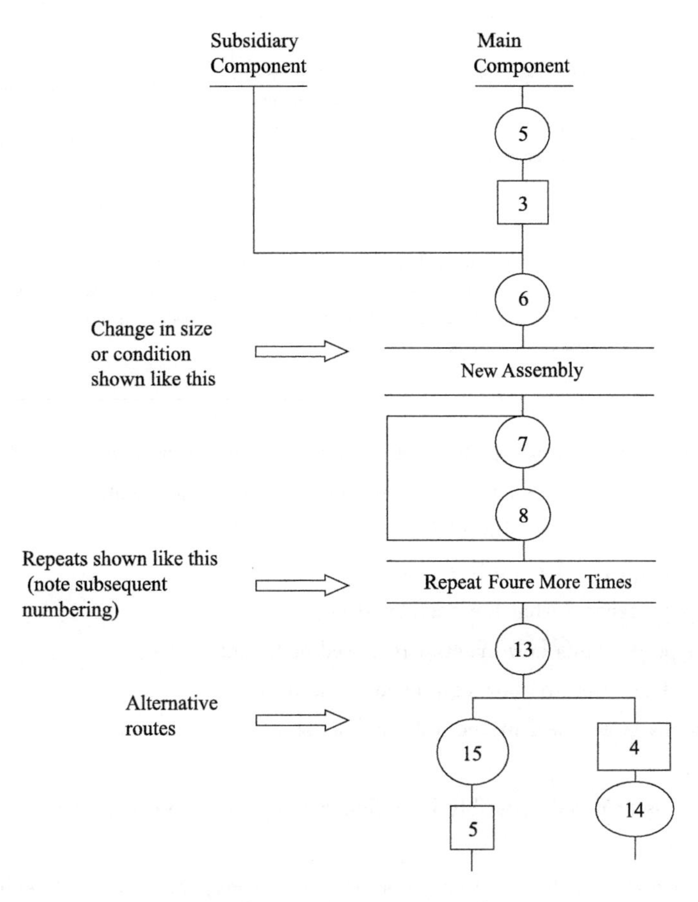

Figure 2-2 Operation Process Chart

Table 2-3 Five Symbols That Present Flow Process Chart

Activity Symbols	Explain	Example
○ Operation	Indicates the main steps in a process, method or procedure. Usually the part, material or product concerned is modified or changed during the operation	Drive nail, drill hole, type letter, etc.
⇨ Transport	Indicates the movement of workers, materials or equipment from place to place	Move material by truck, elevator or carrying
□ Inspection	Indicates an inspection for quality and / or a check for quantity	Examine material for quality or quantity

(Contained)

Activity Symbols	Explain	Example
D Temporary Storage or Delay	Indicates a delay in the sequence of events: for example, work waiting between consecutive operations, or any object laid aside temporarily without record until required	Material in truck or on floor at bench waiting to be processed Employee waiting for elevator
▽ Storage	Indicates a controlled storage in which material is received into of issued from a store under some form of authorization, of an item retained for reference purposes	Bulk storage of raw material, finished product in warehouse

A flow process chart is a process chart setting out the sequence of flow of a product or a procedure by recording all events under review with the appropriate process chart symbols. This chart expands the version of operation process chart with all five ASME symbols.

There are three types of flow process charts:

- Worker type—records what the operator does;
- Material type—records how material is moved or treated;
- Equipment type—records how equipment is used.

Flow process charts are used by the following six steps:

(1) Selecting

Selecting a job is analyzed, possibly identified from operation process charts.

(2) Recording

Inspections, operations, delays, transports, storage, times, distances and sequence of activities are recorded on a flow process chart.

(3) Examining

The critical examination techniques are used to eliminate or combine operations or reorganize the sequence of operations.

The additional techniques are possibly decided to use, for example, flow diagrams are used to indicate the more information.

(4) Developing

The information and knowledge gained during recording and examining stages are used to develop a new sequence of operations.

A flow process chart of the new sequence is produced.

(5) Installing

New flow process charts are used to explain proposed changes to managers, supervisors and operators—one of old sequences is compared with one of news.

(6) Maintaining

A new flow process chart is keeping on working, new problems are found and a new turn improvement is done.

Example 1: A product manufacturing flow process chart. Can you point out the unreasonable processes of this chart and improve them?

Job name:	No.:
Start:	End:
Method:	
Reseracher:	Date:
Reviewer	Date:

Statistics			
Item	Number	Time/mir	Distance/m
Operation ○	3	125	
Inspection □	2	16	
Transport →	5	34	240
Temporary storage or delay D	1	60	
Storage ▽	1	60	

Job description	Equipment	time/min	Number	Distance /m	Process series ○	□	→	D	▽
1.Blanking	cutting machine	100	2		●				
2.Move to the next process	forklif	10	1	100			●		
3.Determination of dimensions	vernier caliper	10	1			●			
4.Temporary placement	tray	60	1					●	
5.Move to the next process	forklif	8	1	50			●		
6.Rough turning	lathe	20	1		●				
7.Move to the next process	forklif	7	1	40			●		
8.Precision turing	lathe	5	1		●				
9.Move to the next process	forklif	5	1	30			●		
10.Inspection	vernier caliper	6	1			●			
11.Move to the next process	forklif	4	1	20			●		
12.Storaget	warehouse	60	1						●

Flow Diagramand String Diagram

Flow Diagram

Flow Diagram

A diagram or model, substantially to scale, shows the location of specific activities carried out and the routes followed by workers, materials or equipment in their execution. In the flow diagram, layout of work area and movements are recorded, and the excessive distances, convoluted routes and non-value-added activities are indicated. Then the rearrangement of work area can be done to reduce the traveling distances and non-value-added activities.

String Diagram

The string diagram is a scale plan or model on which a thread is used to trace and measure the path of worker, material or equipment during a specified sequence of events. In the string diagram, the distances moved by workers, material or equipment are recorded, and he excessive distances, convoluted routes are indicated. Then the rearrangement of work area can be done to reduce the traveling distance.

Management Transaction Analysis

With the deepening of the application in industrial engineering methods, the process analysis can also be used in the daily work on the affairs of business, management aspects to improve production efficiency, reduce cost, called management transaction analysis, and using management transaction flow diagram as analytical tools for recording and analysis. With the management transaction analysis can make the process of management scientific, management work standard and automation. Because the management transaction analysis is to convey the information as the main objective, not one person can complete the operation along, so it has the characteristics of involving many staffs and more jobs.

Analysis of symbols used to management transaction flow diagram as shown in Table 2-4.

Table 2-4 The symbols of management transaction flow diagram

Activity Symbols	Meaning
○ Operation	Signed, examined and approved, handling.
⇨ Transport	Indicates the movement of objects or vouchers from place to place.
□ Inspection	Indicates a check for quantity.
◇ Inspection	Indicates an inspection for quality.
⊠ Inspection	Indicates to check the quality of the main, also check the quality.
▱ Voucher	Fill or generate various vouchers or documents.
⬭ Purchased material	Purchase of goods from outside the unit.

Operation Analysis

Operation analysis is used to study all productive and nonproductive activities within one operation at one place. Through studying elements of an operation, its purposes are to increase productivity per unit of time and reduce unit costs while maintaining or improving quality. The difference between process analysis and operation analysis is that process analysis studies the whole production process and the end point is one operation while operation analysis focuses on one operational process and the end point is one activity within the operation. Operation analysis can be classified man-machine operation analysis, gang process analysis and two-hand operation analysis.

Primary Demand of Operation Analysis

(1) Reducing the quantities of operations to the lowest, arranging the processes with reasonable sequence and making every operation easiest by using the principles of ECRS.

(2) Exerting the function of two hands, balancing the burthen of two hands and avoiding holding a workpiece for a long time and making use of tools to the utmost extent.

(3) Letting the machines do the most work.

(4) Reducing the operational cycles and frequencies, decreasing the transportation and transfer times of materials and cutting down the distance of transportation and moving, making the moving and transportation easily.

(5) Eliminating the unreasonable space and giving the worker enough workspace.

(6) Eliminating the unreasonable idleness time and realizing the synchronization work of the machine and operator.

In a word, the objective of operation analysis is to arrange the reasonable work configuration, to reduce the labor strength of operators and to decrease the work time.

Man-Machine Operation Analysis

The man-machine operation analysis is used to investigate, analyze the relationship of worker and machine in the operation cycle. These facts can lead to a fuller utilization of both worker and machine time, and a better balance of the work cycle.

Man-machine operation analysis uses the man and machine chart to show the exact time relationship between the working cycle of the operator and the operating cycle of the machine. So through analyzing the chart we can get new methods to reduce the idle time of worker and machine and to enhance the efficiency of the man-machine system.

The relationship between workers and machines is usually one of three types: synchronous servicing, completely random servicing and a combination of synchronous and random servicing.

Example 2: One worker operates two semi-automatic lathes, the process: loading material 0.5min; turning 1min; returning material 0.25min, two machines processing the same parts, automatically turning and stopping, the time of people coming from one machine to another needs 6 seconds. Try to draw a program chart of the best utilization of this man-machine.

People	Time	Machinel	Time	Machine2	Time
Description		Description		Description	
Returning material of machine 1	0.25	Returned material	0.25	Turning	1
Loading material of machine 2	0.5	Loaded material	0.5		
Go to machinc 2	0.1	Turning	1		
Leisure	0.15				
Return material of machine 2	0.25			Returned material	0.25
Loading material ofmachine 2	0.5			Loaded material	0.5
Efficiency of worker	91.4%	Efficiency of machine 1	100%	Efficiency of machine 2	100%

Construction of the Man-Machine Chart

The chart is composed of two parts. One part can be called the chart title such as man-machine process chart and other information including part number, drawing number, operation description, present or proposed method, date and name of the researcher in the top of the chart. And the other part can be divided into two sides. The left side shows the operations and time for the worker and the right shows the working time and the idle time of the machine or machines.

Gang Process Analysis

In the workplace, there're always two or more than two workers operating with one machine or job, and then this operation is called gang process analysis. The gang process chart shows the exact relationship between the idle and operating cycle of the machine and the idle and operating time per cycle of the workers who serves that machine. This chart presents the possibilities for improvement by reducing both idle operator and machine time.

Two-Hand Process Analysis

In the workplace, workers' two hands finish many concrete motions. Recording and studying these motions of two hands are called two-hand process analysis. The two-hand process chart is a process chart in which the motions of a worker's hands (or limbs) are recorded in an operating cycle to show the exact relationship between two hands' movement. The purpose of the two-hand process chart is to present a given operation in sufficient detail that the operation can be analyzed and improved. In this part there is an example of two-hand process chart study. With this example you can learn to draw this chart and to master the analysis approach.

Selecting: a task has been selected because it seems to be taking too long time for such a relatively simple activity.

Recording: seen in Figure 2-3.

Examining: the following are pickups with the critical examination technique:

Questions:

1. Why is it necessary to hold the tube in the jig?

2. Why should we cannot the tube be notched while it is being rotated instead of the right hand having to wait?

3. Why does the tube have to be taken out the jig to break it?

4. Why should we pick up and put down the file at the end of each cycle? Can't it be held?

Answers:

1. The tube will always have to be held because the length supported by the jig is short compared with the total length of the tube.

2. There is no reason why the tube cannot be rotated and notched at the same time.

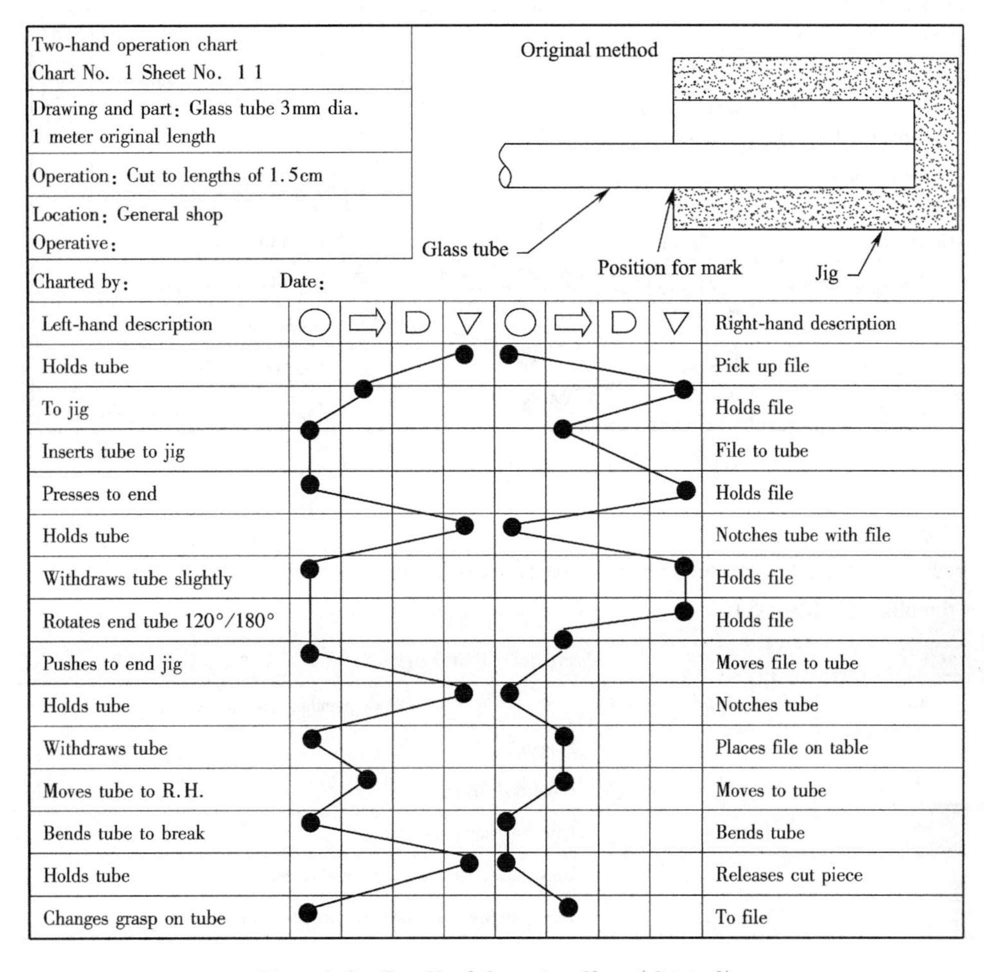

Two-hand operation chart Chart No. 1 Sheet No. 1 1									
Drawing and part: Glass tube 3mm dia. 1 meter original length									
Operation: Cut to lengths of 1.5cm									
Location: General shop Operative:									
Charted by: Date:									
Left-hand description	○	⇨	D	▽	○	⇨	D	▽	Right-hand description
Holds tube				●	●				Pick up file
To jig		●						●	Holds file
Inserts tube to jig	●					●			File to tube
Presses to end	●							●	Holds file
Holds tube				●	●				Notches tube with file
Withdraws tube slightly	●							●	Holds file
Rotates end tube 120°/180°								●	Holds file
Pushes to end jig	●					●			Moves file to tube
Holds tube				●	●				Notches tube
Withdraws tube	●					●			Places file on table
Moves tube to R.H.		●				●			Moves to tube
Bends tube to break	●				●				Bends tube
Holds tube				●	●				Releases cut piece
Changes grasp on tube	●					●			To file

Figure 2-3 Two-Hand Operation Chart (Original)

3. The tube has to be taken out of the jig to be broken because if the tube were broken by bending against the face of the jig, the short end would then have to be picked out—an awkward operation if very little were sticking out; if the jig were so designed that the short end would fall out when broken, it would not then be necessary to withdraw the tube.

4. Both hands are needed to break the tube by using the old method. This might not be necessary if a new jig could be devised.

Motion Analysis

Motion analysis regards one activity as target to analyze the motion in details. The main principle of motion analysis is to analyze, compare and research the motions of the operator's hands, eyes and the other part of body, and then to eliminate the otiose motions, rearrange the necessary and effective motions to a standard set of motions with reasonable tools and layout.

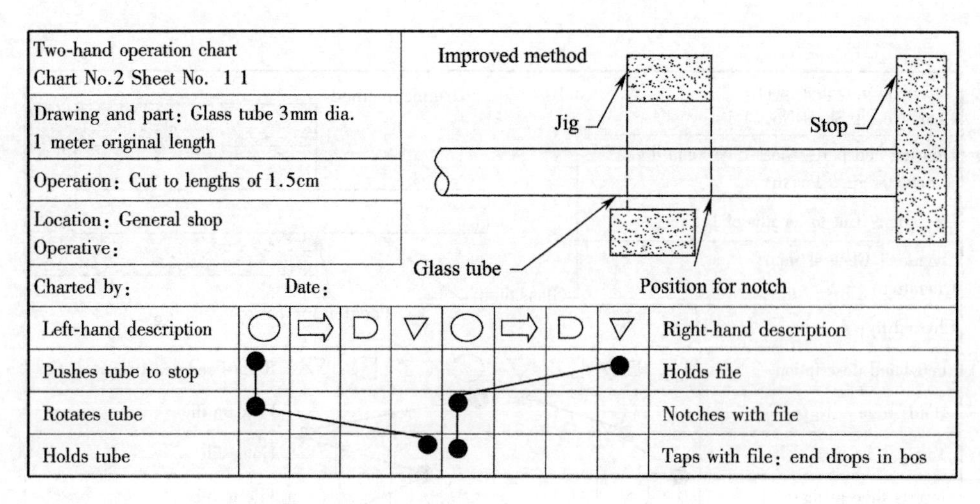

Figure 2-4 Two-Hand Operation Chart (Updated)

Threbligs

F. B. Gilbreth, the creator of motion study, divided the operations of man into 17 smallest units—threbligs (Table 2-5).

Table 2-5 Threbligs

Class	Pivot	Body member (s) moved
1	Knuckle	Fingers
2	Wrist	Hand and fingers
3	Elbow	Forearm, hand and fingers
4	Shoulder	Upper arm, forearm, hand and fingers
5	Trunk	Torso, upper arm, forearm, hand and fingers

Principles of Motion Economy

Principles Related to the Use of Human Body

(1) The two hands should begin and complete their movements at the same time.

(2) The two hands should not be idle at the same time except during periods of rest.

(3) The motions of the arms should be symmetrical and in opposite directions and should be made simultaneously.

(4) Hand and body motions should be made at the lowest level of classifications at which it is possible to work satisfactorily.

(5) Momentum of objects should be made use of to the most extent to help the worker. When the muscle strength is required it should be reduced to the minimum extent.

(6) Continuous curved movements are to be preferred compared with straight-line motions involving sudden and sharp changes direction.

(7) "Ballistic" (i. e. free swinging) movements are faster, easier and more accurate than re-

stricted or controlled movements.[2]

(8) Rhythm is essential to the smooth and automatic performance of a repetitive operation. The work should be arranged to permit easy and natural with them whenever possible.

(9) Work should be arranged so that eye movements are confined to a comfortable area, without the need for frequent changes of focus.

Principles Related to the Arrangement of Work Place

(1) Definite and fixed stations should be provided for all tools and materials.

(2) Tools and materials should be pre-positioned to reduce searching time.

(3) Gravity feed, bins and containers should be used to deliver materials as close to the point of use as possible.

(4) Tools, materials and controls should be located within the maximum working area and as near to the worker as possible.

(5) Materials and tools should be arranged to operate the best sequence of motions.

(6) "Drop deliveries" or ejectors should be used wherever possible, so that the operator does not use his or her hands to dispose the finished goods.

(7) Workplace should have adequate lighting and the type and height of a chair make operator have good posture and feel comfortable. The height of working-table or of a chair should be adjusted to suit alternate standing and sitting.

(8) The colors in the workplace should harmonize with that kind of the work and thus eye fatigue is reduced and operators feel comfortable.

Principles Related to Tools and Equipment

(1) The hands should be relieved of all work of "holding" the work piece where work can be done by a jig, fixture or foot operated device.

(2) Two or more tools should be combined wherever possible.

(3) Where each finger performs some specific movement, as in typewriting, the load should be distributed in accordance with the inherent capacities of the fingers.

(4) Handles, such as cranks and large screwdrivers, should be designed in this way that much of the surface of the handle can be contacted by hands. This is especially necessary when considerable force has to be put on the handle.

(5) Levers, crossbars and hand wheels on machinery should be placed in such positions in order that the operator can use them with the least change of body position and can make use of the greatest "mechanical capabilities".

Notes

1. The work study is an integrative technology of engineering and management which can increase the ratio of output without investment or with little investment, so it has been paid the general attention of the industrial circle for a long time.

句意：方法研究是一项集工程与管理于一体的技术，其目的是寻求在不投资或少投资的情况下来提高生产效率，因而多年来在工业界一直受到普遍的关注。

2. "Ballistic" (i. e. free swinging) movements are faster, easier and more accurate

than restricted or controlled movements.

句意：自然带弧形的运动比受限制或受控制的运动轻快、准确。

Exercise

1. What are the 5W1H technologies of process analysis?
2. What are the "ECRS" principles of process analysis?
3. What are the principles of motion economy?
4. Recite the definition of the process analysis.
5. Describe the drawing procedure of the outline process chart with your own words.
6. According to your own practice, drawing the flow process chart and flow diagram of those things.

(1) Washing the clothes by washer, and then drying them on the bamboo.

(2) Writing a letter and mailing it in the nearby pillar box.

Unit 2 Time Study

What Is Time Study?

Time study is one of the work measurement techniques. [1] Time study is a structured process of directly observing and measuring (using a timing device) human work in order to establish the time required for completion of the work by a qualified worker when working at a defined level of performance. [2] Here, "a qualified worker" means that the worker is suitable for that job (or task), he trained to know how to do that job, and he has normally physiological characteristics and works at the normal speed. "Working at a defined level of performance" indicates that the work methods, equipment, procedures, motions, tools, rotate speed of the machine and work environment are standardized.

The Methods of Determining Time Standards

Three methods help determine time standards: estimates, historical records, and work measurement procedures.

In past years, researchers relied more heavily on estimates as a means of establishing standards. With today's increasing competition from foreign producers, there has been an increasing effort to establish standards based on facts rather than judgment. Experience has shown that no individual can establish consistent and fair standards simply by looking at a job and judging the amount of time required to complete it. Where estimates are used, standards are out of line. Compensating errors sometimes diminish this deviation, but experience shows that over a period of time, estimated values deviate substantially from measured standards. Both historical records and work measurement techniques give much more accurate values than the use of estimates based on judgment alone.

With the historical records method, production standards are based on the records of similar previously performed jobs. In common practice, the worker punches in on a time clock or data collection hardware every time he or she begins a new job, and then punches out after completing the job. This technique tells how long it actually took to do a job, but not how long it should have taken.[3] Since operators wish to justify their entire working day, some jobs carry personal, unavoidable, and avoidable delay time to a much greater extent than they should, while other jobs do not carry their appropriate share of delay time. Historical records have consistently deviated by as much as 50 percent on the same operation of the same job. Yet, as a basis of determining labor standards, historical records are better than no records at all. Such records give more reliable results than estimates based on judgment alone, but they do not provide sufficiently valid results to assure equitable and competitive labor costs.

Any of the work-measurement techniques—stopwatch (electronic or mechanical) time study, fundamental motion data, standard data, time formulas, or work sampling studies—represent a better way to establish fair production standards. All of these techniques are based on facts. They establish an allowed time standard for performing a given task, with due allowance for fatigue, for policies, for specials, and for personal and unavoidable delays.

The Significance and Purposes of Time Study

Accurately established time standards make it possible to produce more within a given plant, thus increasing the efficiency of the equipment and the operating personnel. Poorly established standards, although better than no standards at all, lead to high costs, labor dissension, and possibly even the failure of the enterprise. Sound standards will mean the difference between the success or failure of a business.

Time study has the following purposes:

To determine the time standards and control labor costs.

To set up the standard times and take them as the base of payroll.

To decide the work schedule and work plan.

To determine the standard cost and take it as the evidence of standard budget.

To determine the utilization efficiency of machines and to solve the problems on balance of production lines.

Requirements of Time Study

Certain fundamental requirements must be realized before the time study is taken. For example, whether the standard is required on a new job, or on an old job in which the method or part of the method has been altered, the operator should be thoroughly acquainted with the new technique before the operation is studied. Also, the method must be standardized at all points where it is to be used, before the study begins. Unless all details of the method and working conditions have been standardized, the time standards will have little value and will become a continual source of mistrust, grievances, and internal friction.

Researchers of time study should tell all related people including the union steward, the department supervisor, and the operator that the job is to be studied. Each of these parties can then make specific advanced plans and take the steps necessary to allow a smooth, coordinated study. The operator should verify that he or she is performing the correct method and should become acquainted with all details of that operation. The supervisor should check the method to make sure that feeds, speeds, cutting tools, lubricants, and so forth conform to standard practice, as established by the methods department. Also, the supervisor should investigate the amount of material available so that no shortages take place during the study. If several operators are available for the study, the supervisor should determine which operator would give the most satisfactory results. The union steward should then make sure that only trained, competent operators are selected, should explain why the study is being taken, and should answer any pertinent questions raised by the operator.

Tools of Time Study

The minimum tools required to conduct a time study program include a stopwatch, time study board, time study forms, and pocket calculator. Videotape equipment can also be very useful.

Stopwatch

Two types of stopwatches are in use today: ① the traditional decimal minute watch (Figure 2-5) (0.01 minute), and ② the much more practical electronic stopwatch (Figure 2-6). The decimal minute watch, shown in Figure 2-5, has 100 divisions on its face, and each division is equal to 0.01 minute, that is, a complete sweep of the long hand requires one minute. The small dial on the watch face has 30 divisions, each of which is equal to one minute. Therefore, for every full revolution of the sweep hand, the small hand moves one division, or one minute. To start this watch, move the side slide toward the crown. Moving the side slide away from the crown stops the watch with the hands in their existing positions. To continue operation of the watch from the point where the hands stopped, move the slide toward the crown. Depressing the crown moves both the sweep hand and the small hand back to zero. Releasing the crown puts the watch back into operation, unless the side slide is moved away from the crown.

Figure 2-5　A Decimal Minute Watch

These watches provide resolution to 0.001 second and an accuracy of ±0.002 percent. They permit timing any number of individual elements, while also counting the total elapsed time. Thus, they provide both continuous and snapback timing (button C), with none of the disadvantages of mechanical watches. To operate the watch, press the top button

(button A). Each time the top button is pressed, a numerical readout is presented. Pressing the memory button (button B) causes previous readouts to be retrieved. A slightly fancier version incorporates the watch into an electronic time study board (Figure 2-7).

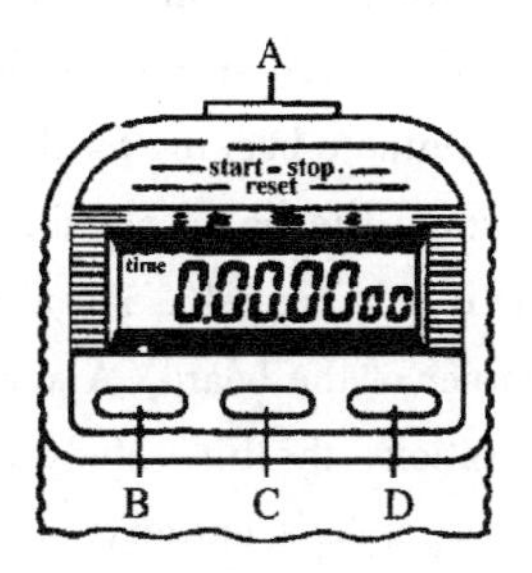

Figure 2-6 An Electronic Stopwatch

A—Start/stop B—memory retrieval, C—mode (continuous/snapback), and D—other functions

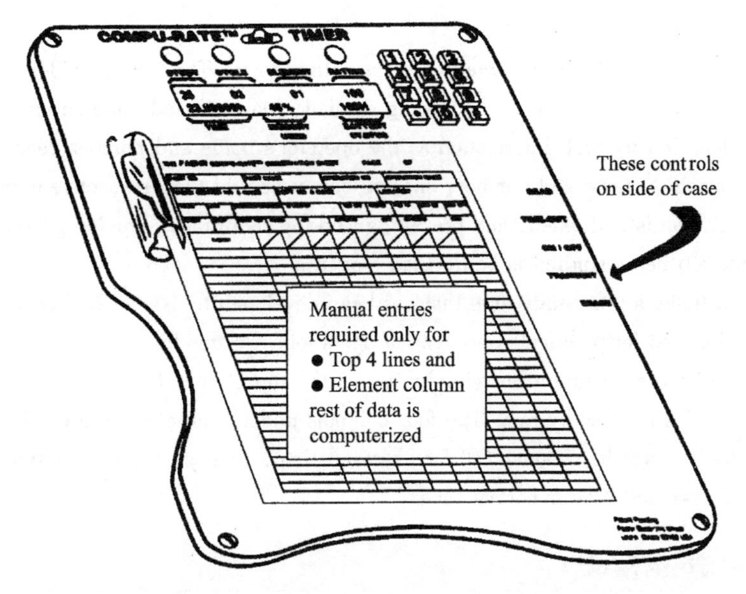

Figure 2-7 A Computer-Assisted Electronic Stopwatch

Videotape Cameras

Videotape cameras are ideal for recording operators' methods and elapsed time. By taking pictures of the operation and then studying them a frame at a time, researchers can record exact details of the method used and can then assign normal time values. They can also establish standards by projecting the film at the same speed that the pictures were taken and then performance rating the operator. Because all the facts are there, observing the videotape is a fair and accurate way to rate performance. Then, too, potential methods improvements that would seldom be uncovered with a stopwatch procedure can be revealed through the camera eye.

Time Study Board

When the stopwatch is being used, researchers find it convenient to have a suitable board to hold the time study form and the stopwatch. The board should be light, so as not to tire the arm, and yet strong and sufficiently hard to provide a suitable backing for the time study form. Suitable materials include 1/4-inch plywood or smooth plastic. The board should have both arm and body contacts, for comfortable fit and ease of writing while it is being held. For a right-handed observer, the watch should be mounted in the upper right-hand corner of the board. A spring clip to the left would hold the time study form. [4] Standing in the proper position, the time study researcher can look over the top of the watch to the workstation and follow the operator's movements, while keeping both the watch and the time study form in the immediate field of vision.

Time Study Forms

All the details of the study are recorded on a time study form. The form provides space to record all related information concerning the method being studied, tools utilized, and so on. The operation being studied is identified by such information as the operator's name and number, operation description and number, machine name and number, special tools used and their respective numbers, department where the operation is performed, and prevailing working conditions. Providing too much information concerning the job being studied is better than too little.

Table 2-4 illustrates a time study form that has been developed by Niebel & Freivalds. It is sufficiently flexible to be used for practically any type of operation. On this form, researchers would record the various elements of the operation horizontally across the top of the sheet, and the cycles studied would be entered vertically, row by row. The four columns under each element are: R for ratings; W for watch time, that is, watch readout; OT for observed time, that is, the differential time between successive watch times; and NT for normal time.

Process of Time Study

The actual conduct of time study is both a science and art. To ensure success, researchers must be able to inspire confidence in, exercise judgment with, and develop a personable approach to everyone with whom they come in contact. In addition, their backgrounds and training should prepare them to understand thoroughly and perform the various functions related to the study. These elements include selecting the operator, analyzing the job and breaking it down into its elements, recording the elapsed elemental values, performance rating the operator, assigning appropriate allowances, and working up the study itself. [2]

Choosing the Operator

The first step in beginning a time study is carried out by the departmental or line supervisor. Af-

ter reviewing the job in operation, both the supervisor and the time study researcher should agree that the job is ready to be studied. If more than one operator is performing the work for which the standard is to be established, several things should be considered when selecting which operator to use for the study. In general, an operator who is average or somewhat above average in performance gives a more satisfactory study than a low-skilled or highly superior operator. The average operator usually performs the work consistently and systematically. That operator's pace will tend to be approximately in the standard range, thereby making it easier for the time study analyst to apply a correct performance factor.

Recording Significant Information

Researchers should record the machines, hand tools, jigs or fixtures, working conditions, materials, operations, operator name and clock number, department, study date, and observer's name (Table 2-6). Space for such detail is provided under Remarks on the time study form. A sketch of the layout may also be helpful on another page of the time study form. The more pertinent information is recorded, the more useful the time study becomes over the years. It becomes a resource for establishing standard data and developing formulas. It will also be useful for methods improvement, operator evaluation, tool evaluation, and machine performance evaluation.

When machine tools are used, the researcher should specify the name, size, style, capacity, and serial or inventory number, as well as the working conditions. Dies, jigs, gages, and fixtures should be identified by their numbers and with short descriptions. If the working conditions during the study are different from the normal conditions for that job, they will affect the performance of the operator. For example, in a drop forge shop, if a study were taken on an extremely hot day, the working conditions would be poorer than usual, and operator performance would reflect the effect of the intense heat. Consequently, an allowance would be added to the operator's normal time. If the working conditions improve, the allowance can be diminished. Conversely, if the working conditions become poorer, the allowance should be raised.

The operation performed should be very specifically described. For example, "broach 3/8 inch × 3/8 inch keyway in 1-inch bore" is considerably more explicit than "broach keyway." The operator being studied should be identified by name and clock number; there could easily be two John Smiths in one company.

Table 2-6 Time Study Observation Form

Time Study Observation Form						Study No. 2-85				Date: 3-1				Page: 1of 1			
						Operation: Die casting				Operator: B Jones				Observer: A F			
Element No. & Description		1 Remove part from die, lubricate die, inspect				2 Race part in fixture, trim as die part				3				4			
Note	Cycle	R	W	OT	NT	R	W	OT	NT	R	W	OT	NT	R	W	OT	NT
	1	90	90	30	270	90	113	23	207								
	2	100	40	27	270	100	61	21	210								

(Contained)

Note	Cycle	R	W	OT	NT	R	W	OT	NT	R	W	OT	NT	R	W	OT	NT
	3	90	92	31	279	90	215	23	207								
	4	85	50	35	298	100	70	20	200								
	5	100	98	28	280	100	318	20	200								
	6	110	43	25	275	110	61	18	198								
	7	90	92	31	279	90	416	24	216								
	8	100	44	28	280	85	68	24	204								
	9	90	500	32	288	90	23	23	207								
	10	110	49	26	286	105	68	19	200								

Summary				
Total OT	2.93	2.15		
Rating	—	—		
Total NT	2.805	2.049		
No. Observations	10	10		
Average NT	0.281	0.205		
Allowance (%)	17	17		
Element Std. Time	0.329	0.240		
No. Occurrence	1	1		
Standard time	0.329	0.240		
Total standard time (sum standard time for all elements)				0.569

Foreign Elements					Time Check			Allowance Summary	
Sym.	W1	W2	OT	Des.	Finishing Time		3:48	Personal Needs	5
A					Starting Time		3:42	Basic Fatigue	4
B					Elapsed Time		6	Variable Fatigue	8
C					TEBS	0.60		Special	—
D					TEAF	0.32		Total Allowance	17%
E					Total Check Time	0.92			
F					Effective Time	5.08		Remarks:	
G					Ineffective Time	0			
Rating Check					Total Recorded Time	6.00			
Synthetic Time		(%)			Unaccounted Time	0			
Observed Time					Recording Error (%)	0			

(Source: Reference [22])

Positioning the Observer

The observer should stand, not sit, a few feet to the rear of the operator, so as not to distract or interfere with the worker. [5] Standing observers are better able to move around and follow the movements of the operator's hands as the operator goes through the work cycle. During the course of the study, the observer should avoid any conversation with the operator, as this could distract the worker or upset the routines.

Dividing the Operation into Elements

For ease of measurement, the operation should be divided into groups of motions known as elements. To divide the operation into its individual elements, the researcher should watch the operator for several cycles. [6] However, if the cycle time is over 30 minutes, the researcher can write the description of the elements while taking the study. If possible, the researcher should determine the operational elements before the start of the study. Elements should be broken down into divisions that are as fine as possible and yet not so small that reading accuracy is sacrificed. Elemental divisions of around 0.04 minute are about as fine as can be read consistently by an experienced time study researcher. However, if the preceding and succeeding elements are relatively long, an element as short as 0.02 minute can be readily timed.

To identify endpoints completely and develop consistency in reading the watch from one cycle to the next, consider both sound and sight in the elemental breakdown. For example, the breakpoints of elements can be associated with such sounds as a finished piece hitting the container, a facing tool biting into a casting, a drill breaking through the part being drilled, and a pair of micrometers being laid on a bench.

Each element should be recorded in its proper sequence, including a basic division of work terminated by a distinctive sound or motion. For example, the element "up part to manual chuck and tighten" would include the following basic divisions: reach for part, grasp part, move part, position part, reach for chuck wrench, grasp chuck wrench, move chuck wrench, position chuck wrench, turn chuck wrench, and release chuck wrench. The termination point of this element would be the chuck wrench being dropped on the head of the lathe, as evidenced by the accompanying sound. The element "start machine" could include reach for lever, grasp lever, move lever, and release lever. The rotation of the machine, with the accompanying sound, would identify the termination point so that readings could be made at exactly the same point in each cycle.

Frequently, different time study researchers in a company adopt a standard elemental breakdown for given classes of facilities, to ensure uniformity in establishing breakpoints. For example, all single-spindle bench-type drill press work may be broken down into standard elements, and all lathe work may be composed of a series of predetermined elements. Having standard elements as a basis for operation breakdown is especially important in the establishment of standard data.

Some additional suggestions that may help in breaking elements down are:

(1) In general, keep manual and machine elements separate, since machine time is less affected by ratings.

(2) Likewise, separate constant elements (those elements for which the time does not deviate within a specified range of work) from variable elements (those elements for which the time does vary within a specified range of work).

(3) When an element is repeated, do not include a second description. Instead, in the space provided for the element description, give the identifying number that was used when the element first occurred.

Rating[7]

Since the actual time required to perform each element of the study depends on a high degree on the skill and effort of the operator, it is necessary to adjust upward the time of good operator and the time of the poor operator downward to a standard level. On short-cycle, repetitive work, it is customary to apply one rating to the entire study, or an average rating for each element. However, when the elements are long and entail diversified manual movements, it is more practical to evaluate the performance of each element as it occurs.

Allowances[8]

No operator can maintain a standard pace every minute of the working day. Five classes of interruptions can take place, for which extra time must be provided. The first is personal interruptions, such as trips to the restroom and drinking fountain; the second is fatigue, which can affect even the strongest individual on the lightest work. Thirdly, there are unavoidable delays, such as tool breakage, supervisor interruptions, slight tool trouble, and material variations. The fourth is special needs, such as grinding tools, clearing machines, operating multi-machines, and dealing with other temporary happened events. Finally, policy allowance is necessary for management needs. All of which require that some allowance be made.

The Standard Time

The time required for a fully qualified, trained operator, working at a standard pace and exerting average effort, to perform the operation is termed the standard time (ST) for that operation.

The standard time is calculated by the following equations:

$$NT = OT \times R/100 \tag{2-1}$$

where R is the performance rating of the operator expressed as a percentage, with 100 percent being standard performance by a qualified operator. To do a fair job of rating, the time study researcher must be able to disregard personalities and other varying factors, and consider only the amount of work being done per unit of time, as compared to the amount of work that the qualified operator would produce.

$$ST = NT + NT \times Allowance = NT \times (1 + Allowance) \tag{2-2}$$

Notes

1. Time study is one of the work measurement techniques.

句意：时间研究是一种作业测定技术。

2. Time study is a structured process of directly observing and measuring (using a timing device) human work in order to establish the time required for completion of the work by a qualified worker when working at a defined level of performance.

句意：时间研究是一种结构化的直接观察和测量（使用计时设备）过程，旨在决定一位合格、适当、训练有素的操作者，在标准状态下，对一特定的工作以正常速度操作所需要的时间。

3. This technique tells how long it actually took to do a job, but not how long it should have taken.

句意：这一技术说明做一件工作实际花多长时间，而不是它应该花多长时间。

4. A spring clip to the left would hold the time study form.

句意：在左边的夹子应能夹住时间研究表。

5. The observer should stand, not sit, a few feet to the rear of the operator, so as not to distract or interfere with the worker.

句意：观察者应该站着而不是坐着，在操作者背后几英尺远的地方，不要转移操作者的注意力或者干扰操作者操作。

6. To divide the operation into its individual elements, the researcher should watch the operator for several cycles.

句意：为了划分操作单元，研究者应该观察操作者几个操作循环。

7. rating 评比

8. allowance 宽放

Exercises

1. What are the effects of poor time standards?

2. What equipment is needed by the time study researchers?

3. The time study researcher at the Dongfang Company developed for the following snapback stopwatch readings where elemental performance rating was used. The allowance for this element was assigned a value of 12 percent. What would be the standard time for this element?

4. The following data came from a time study taken on a horizontal milling machine:

Mean manual effort time per cycle: 4.82 minutes.

Mean cutting time (power feed): 3.54 minutes.

Mean performance rating: 120 percent.

Machine allowance (power feed): 11 percent.

5. Define a qualified operator.

6. Why are allowances applied to the normal time? How many allowances should be considered when setting up standard time?

Unit 3 Time Measurement Methods

Direct Time Study

Direct time study is a method of using time measurement to determine the standard time.

At the start of the study, record the time of day (on a whole minute) from a "master" clock while simultaneously starting the stopwatch. (It is assumed that all data are recorded on the time study form.) This is the starting time as shown in Table 2-7. One of two techniques can be used for recording the elemental times during the study. The one is the continuous timing method. As the name implies, it allows the stopwatch to run for the entire duration of the study. In this method, the researcher reads the watch at the breakpoint of each element, and the time is allowed to continue. The other is the snapback technique. It means that after the watch is read at the breakpoint of each element, the watch time is returned to zero; as the next element takes place, the time increments from zero.

When recording the watch readings, note only the necessary digits and omit the decimal point, thus giving as much time as possible to observing the performance of the operator. [1] If using a decimal minute watch, if the breakpoint of the first element occurs at 0.08 minute, record only the digit 8 in the W (watch time) column. Other example recordings are shown in Table 2-7.

Table 2-7 Summary of Steps in Performing and Computing a Derect Time Study

Time Study Observation Form						Study No. 1-3				Date: 3-22				Page: 1of 1			
						Operation: Machining				Operator: J Smith				Observer: A F			
Element No. & Description		1 Feed bar to stop				2 Index, feed cutting tool to bar				3 Turn 1/2" 550rpm				4 With draw tool and bar set down			
Note	Cycle	R	W	OT	NT	R	W	OT	NT	R	W	OT	NT	R	W	OT	N'
	1	85		19	162	105		12	126	100		60	600	90		17	15
	2	90		20	198	105		13	137	100		60	600	100		16	16
	3	100		17	170	105		11	116	100		60	600	105		17	17
	4																
	5																
	6																
	7																
	8																
Summary																	
Total OT		0.58				0.36				1.80				0.50			
Rating		—				—				—				—			
Total NT		0.530				0.379				1.800				0.492			

Table 2-7 Summary of Steps in Performing and Computing a Derect Time Study (Contained)

No. Observations	3	3	3	3
Average NT	0. 177	0. 126	0. 600	0. 164
Allowance (%)	10	10	10	10
Element Std. Time	0. 195	0. 139	0. 660	0. 180
No. Occurrence	1	1	1	1
Standard Time	0. 195	0. 139	0. 660	0. 180
Total Standard Time (sum standard time for all elements)				1. 174

Foreign Elements					Time Check		Allowance Summary	
Sym.	W1	W2	OT	Descripti On	Finishing Time	9:22	Personal Needs	5
A	0	35	35	Check Dim.	Starting Time	9:16	Basic Fatigue	4
B					Elapsed Time	6	Variable Fatigue	1
C					TEBS	1. 86	Special	—
D					TEAF	0. 60	Total Allowance	10%
E					Total Check Time	2. 46	Remarks: Machine Cycle (element #3) Time = 0. 60 min.	
F					Effective Time	3. 24		
G					Ineffective Time	0. 35		
Rating Check					Total Recorded Time	6. 05		
Synthetic Time	(%)				Unaccounted Time	0. 05		
Observed time					Recording Error	0. 8%		

Source: Reference[22]

The Continuous Timing Method

The continuous method of recording elemental values is superior to the snapback method for several reasons. The most significant is that the resulting study presents a complete record of the entire observation period; as a result, it appeals to the operator and the union. The operator is able to see that no time has been left out of the study, and all delays and foreign elements have been recorded. Since all the facts are clearly presented, this technique of recording times is easier to explain and sell.

The continuous method is also better adapted to measuring and recording very short elements. With practice, a good time study analyst can accurately catch three successive short elements (less than 0. 04 minute), if they are followed by an element of about 0. 15 minute or longer. This is possible by remembering the watch readings of the breakpoints of the three short elements and then recording their respective values while the fourth, longer element is taking place.

On the other hand, more clerical work is involved in calculating the study if the continuous method is used. Since the watch is read at the breakpoint of each element while the hands of the watch continue their movements, it is necessary to make successive subtractions of the consecutive readings to determine the elapsed elemental times. For example, the following readings might represent the breakpoints of a 10-element study: 4, 14, 19, 121, 25, 52, 61, 76, 211, 16. The elemental values of this cycle would be 4, 10, 5, 102, 4, 27, 9, 15, 35, and 5. Table 2-8 presents an example of using continuous timing method.

Finally, we want to explain what are "foreign elements"? During a time study, the operator may encounter unavoidable delays, such as an interruption by a supervisor or tool breakage. The operator may intentionally cause a change in the order of work by going for a drink of water or stopping to rest. These interruptions are referred to as "foreign elements."

Table 2-8 Recordings of Time Study

	①		②		③		④		⑤		Foreign Elements			
Cycle	R	T	R	T	R	T	R	T	R	T	Symbol	R	T	Remarks
1	13	13	28	15	53	25	×		66		A	$\frac{85}{63}$	33	Change Belts
2	84	18	104	20	27	23	39	12	/		B	$\frac{425}{94}$	31	Change & Adjust Screws
3	54	15	72	18	$\frac{205}{85}$	20	$\frac{85}{72}$	14	222	17	C	—		Tools Left and Adjust
4	36	14	53	17	306	A 20	20	14	38	18	D	—		
5	52	14	68	16	87	19	431	B 6	49	18	E	—		
6	64	15	81	17	501	20	23	C 22	41	18	F	—		
7											G	—		

Snapback Method

The snapback method has both advantages and disadvantages compared to the continuous technique. Sometimes study researchers use both methods, believing that studies of predominantly long elements are more adapted to snapback readings, while short-cycle studies are better suited to the continuous method. [2]

Since elapsed element values are read directly in the snapback method, no clerical time is needed to make successive subtractions, as for the continuous method. Thus, the readout can be inserted directly in the OT (observed time) column. Also, elements performed out of order by the operator can be readily recorded without special notation. In addition, proponents of the snapback method state that delays are not recorded. Also, since elemental values can be compared from one cycle to the next, a decision could be made as to the number of cycles to study. However, it is actually erroneous to use observations of the past few cycles to determine how many additional cycles to study. This practice can lead to studying entirely too small a sample.

Among the disadvantages of the snapback method is that it encourages the removal of individual elements from the operation. These cannot be studied independently, because elemental times depend

on the preceding and succeeding elements. Consequently, omitting such factors as delays, foreign elements, and transposed elements could allow erroneous values in the readings accepted. One of the main objections to the snapback method is the amount of time lost while snapping the hand back to zero. This can be anywhere from 0.0018 to 0.0058 minutes. However, this has been negated by the use of electronic watches, for which no time is lost in resetting the readout to zero. Also, short elements (0.04 minute and less) are more difficult to time with this method. Finally, the overall time must be verified by summing the elemental watch readings, a process that is more prone to error.

Cycles in Study

Determining how many cycles to study to arrive at an equitable standard is a subject that has caused considerable discussion among time study researchers, as well as union representatives. Since the activity of the job, as well as its cycle time, directly influences the number of cycles that can be studied from an economic standpoint, the researcher cannot be completely governed by sound statistical practice that demands a certain sample size based on the dispersion of individual element readings. The General Electric Company has established Table 2-9 at an approximate guide to the number of cycles to observe.

Table 2-9 Recommended Number of Observation Cycles

Cycle Time in Minutes	Recommended Number of Cycles
0.10	200
0.25	100
0.50	60
0.75	40
1.00	30
2.00	20
2.00 – 5.00	10
5.00 – 10.00	8
10.00 – 20.00	5
20.00 – 40.00	5
40.00 – above	3

A more accurate number can be established using statistic methods. Since time study is a sampling procedure, the observations can be assumed to be distributed normally about an unknown population mean with an unknown variance. Using the following equation can calculate the number of cycles when the mean error equals to ±5% and reliability equals to 95% according to the principle of statistics:

$$N = \left[40\sqrt{n\sum_{i=1}^{n} X_i^2 - (\sum_{i=1}^{n} X_i)^2} / \sum_{i=1}^{n} X_i\right]^2 \tag{2-3}$$

Where N—the number of cycles

n—the number of cycles in pilot study

x_i—the element readings

For example, a pilot study of 10 readings for a given element is as follows:

$$7, 5, 6, 8, 7, 6, 7, 6, 6, 7$$

Suppose the mean error is ±5% and reliability equals to 95%, ask for the number of cycles N.

According to the equation 2-3, first $\sum x_i$ and $(\sum x_i)^2$ are computed and shown in the Table 2-10.

Table 2-10 Results of Computation

i	1	2	3	4	5	6	7	8	9	10	合计
$\sum x_i$	7	5	6	8	7	6	7	6	6	7	65
$(\sum x_i)^2$	49	25	36	64	49	36	49	36	36	49	429

Put the above data into the equation 2-3:

$$N = \left[\frac{40\sqrt{10 \times 429 - (65)^2}}{65}\right]^2 = 24.6 \approx 25 \text{ observations}$$

10 readings have been done. There are 15 observations should be done.

Performance Rating

The basic principle of performance rating is to adjust the mean observed time (OT) for each element performed during the study to the normal time (NT) that would be required by the qualified operator to perform the same work. Based on the equation 2-1, NT = OT × R/100, we can get the normal time.

Adding Allowances

As mentioned in the first section, no operator can maintain a standard pace every minute of the working day. Allowances must be added to the standard time. Since the direct time study is taken over a relatively short period, and since foreign elements should have been removed in determining the normal time, an allowance must be added to the normal time to arrive at a fair standard that can reasonably be achieved by an operator. The allowance is typically given as a fraction of normal time and is used as a multiplier equal to 1 + allowance.

$$ST = NT + NT \times Allowance = NT \times (1 + Allowance)$$

An alternative approach is to formulate the allowances as a fraction of the total workday, since the actual production time might not be known. In that case, the expression for standard time is:

$$ST = NT/(1 - Allowance) \quad (2\text{-}4)$$

Up to now, there are still many companies which use direct time study to determine the standard time for their operations. Here is an example to compute the standard time.

In a unit, the observed time is 0.8 min, R = 110, allowance = 5%,

$$NT = OT \times R/100 = 0.8 \times 110/100 = 0.88 \text{ min.}$$

$$ST = NT \times (1 + \text{Allowance}) = 0.88 \times (1 + 5\%) = 0.924 \text{ min.}$$

Introduction to Predetermined Time Systems

Another industrial engineering pioneer—Frank B. Gilbreth is the inventor of motions study. He classified the basic motions into what he called Therbligs (which is almost Gilbreth spelled backwards) that is the base of predetermined time systems. Since 1945, there has been a growing interest in the use basic motion times as a method establishing rates quickly and accurately without using the direct time study.

An American engineer, A. B. Segur first took the time study into the motion study in 1924. In his paper, *Motion Time Analysis*, he wrote, "the spent time of all the skilled people doing a piece of basic motion was a constant under the specified conditions.[3]"

In 1934, J. H. Quick, worked in the American Radio Company, created work factor systems (WF).

In 1948, N. B. Maynard, G. J. Stegemerten and J. L. Sckwab, worked in the American Westinghouse Electrics Corporation, published their research achievements, *Methods Time Measurement (MTM)*.

In 1966, anAustralian, Dr. G. C. Heyde created Modular Arrangement of Predetermined Time Standard (MOD) through long time study.

The Characteristics of Predetermined Time Systems

(1) Determine the standard time in advance and objectively in work measurements without the performance rating.

(2) Record the operational ways in details and obtain the elapsed time of each basic motion. Therefore, the operational improvement can be done through analysis.

(3) Determine the standard time before the work being done without using the stopwatch. In this way, the operational regulations can be setup.

(4) When the operational ways update, the standard time values must be altered. However, the base of setting up the standard time, predetermined time systems, remain the same.

(5) It is the best way to balance the production lines with predetermined time systems.

There are over 50 predetermined time systems being used at present in the world. Most of them require the specialized training to the practical applications of these techniques. Here, we introduce some typical and popular systems, such as work factors, Methods Time Measurement, and Modular Arrangement of Predetermined Time Standard.

The Method of Work Factors

The method of work factors is the basic one of predetermined time systems. It determines work time based on the motion features and conditions through thespecified time values in advance. In the method of work factors, there are four variables that inflect motion time. ① when motion, which portion of body is working, ② the motion distance, ③ the gravitation or resistance overcome when motion, and ④ adjustment by human beings (including stop, guidance of direction, specially care of, and direction change). Because the factor 3 and factor 4 make the motion delayed, they are called "work factors" that means the difficult extent of motion. These two factors could be represented by the figure of difficulty coefficients. For a basic motion, as long as the portion of body motion, the motion distance, and the number of difficulty coefficients are determined, the time value of this motion could be determined by the tables of work factors. Combining these basic motion time values, the whole operational time could be obtained.

In the method of work factors, there are two methods, respectively called the simple method and the detailed method. In the simple method, time unit is defined as RU. 1 RU equals 0. 001 minute. In the detailed method, time unit is defined as TWU. 1 TWU equals 0. 0001 minute.

In this section, the simple method of work factors is briefly introduced. In the simple method, there are eightbasic motions and five kinds of work factors, which values are determined by the motion difficulties. Eight basic motions are reach, grasp, release, position, assembly, use, disengage, and manipulation. Five kinds of work factors include weights/resistance, show, prudence, update directions, and deadlock. These eight basic motions are not subject to the five kinds of work factors. However, many motions are subject to the five kinds of work factors. When analysis, the influences of the five kinds of work factors must be considered for every motion. The analytical equations are listed. Combining all the motions, the operational standard time could be determined. For every case, the time value could be obtained by the tables of work factors.

Methods Time Measurement (MTM)

Methods Time Measurement gives time values for the fundamental motions of: reach, move, turn, grasp, position, disengage, and release. MTM was defined as "a procedure which analyzes any manual operation or method into the basic motions required to perform it, and assigns each motion a pre-determined time standard which is determined by the nature of the motion and the conditions under which it is made."

In MTM, 1 TMU (time measurement unit) equals 0. 00001 hour. In order to get the standard time values, the researcher summarizes all left-hand and right-hand motions required to perform the job properly. Then, the rated times in TMU for each motion are determined from the methods-time data tables. To determine the time required for anormal performance of the task, the non-limiting motion values are either circled or deleted, as only the limiting motions will be summarized, provided that it is "easy" to perform the two motions simultaneously. For example, if the right hand must reach 20 inches to pick up a nut, the classification would be R20C and the time value would be 19. 8 TMU. If, at the same time, the left hand must reach 10 inches to pick up a cap screw, a designation of R10C with a TMU value of 12. 9 would be in effect. The right hand value would be the limiting value, and the 12. 9 value of the left hand would not be used in calculating the normal time.

The tabulated values do not carry any allowance for personal delays, fatigue, unavoidable delays,

special needs or management needs. When researchers use these values to establish time standards, they must add appropriate allowances to the summary of the synthetic basic motion times.

In MTM, MTM-1 is the basic method. MTM-1 data are the result of frame-by-frame analyses of motion-picture films of diversified areas of work. The data taken from the various films were rated by the Westinghouse techniques, tabulated, and analyzed to determine the degree of difficulty caused by variable characteristics. For example, both the distance and the type of reach affect reach time. Proponents of MTM-1 state that no fatigue allowance is needed in the vast majority of applications, because the MTM-1 values are based on a work rate that can be sustained at steady-state for the working life of a healthy employee.

There are many methods of time measurement, derived from MTM-1, such as MTM-2, MTM-3, MTM-C, MTM-V, MTM-M, MTM-TE, and so forth. They have constituted the system of MTM. The more detailed descriptions could be seen in reference [22].

Modular Arrangement of Predetermined Motion Time Standard (MOD)

MOD is the latest development of predetermined time systems. It is convenient to use, simply and easily to learn, and its computing accuracy is the same as MTM. Therefore, it is widely accepted and used in enterprises.

MOD classifies operational motions into upper limb basic motions and other motions. The statistic data show that over 90 percent motions are carried out by upper limbs. So, it is very important to analyze the motions of upper limbs. MOD focuses on the operational motions. It classifies the operational motions into the upper limb basic motions and the other motions.

There are 21 motions in MOD. The upper limb basic motions have 11. And the other motions have 10. They are presented by a picture of MOD. There are three kinds of motions in the upper limb basic motions including move, grasp, and position. The number of each motion indicates the corresponding time value. 1 MOD equals 0. 129 second. It means that the timeelapsed when the finger moves 2. 5 cm. MOD combines motion symbols and time values into a unit. For example, G3 means that is a complicated grasp motion and its time value is 3 MOD ($3 \times 0.129 = 0.387$s). Using MOD neither needs measuring time nor the performance rate. As long as the motions are analyzed and determined, the standard time could be set up. So, it is a better and more effective way to analyze motions, evaluate methods, set up the standard times, and balance production lines.

Applications of Predetermined Time Systems

There are many applications of predetermined time systems in plants. The essential purpose of predetermined time systems is to decide the standard time. Here is a simple instance to indicate the applications of MOD.

There is a company, which produces the various bushings. The internal diameter of the bushings must be controlled to meet higher accuracy. Therefore, the produced bushings must be measured by the plug gauge[4] (Figure 2-8). There are four updating measurement methods (Figure 2-9). Try to determine the operation time of measuring the internal diameter of the bushings by MOD.

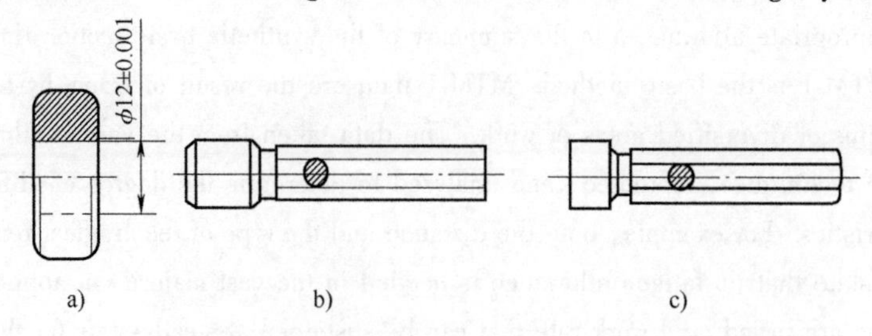

Figure 2-8 A Bushing and Plug Gauges

a) A Bushing b) A plug Gauge (Pass) c) A Plug Gauge (Stop)

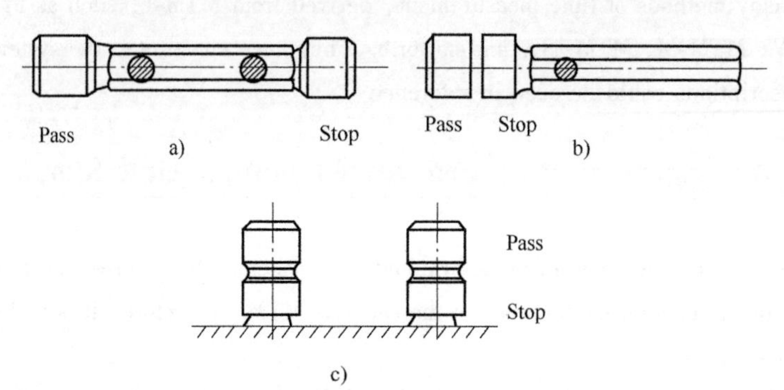

Figure 2-9 Four Updating Measurement Methods

Step 1 Understanding operational procedures

On the site worker's operational procedures are observed. For the first updating situation, the left-hand of the inspector first reaches to the box of bushings, grasps one, and return to the original position. Hold it. The right-hand reaches to the box of the plug gauge, grasps it, and comes back to the original position. The right-hand aims at the hole of the bushing and plugs. Then, the left-hand makes the bushing rotate 180 degrees. The right-hand again aims at the hole of the bushing and plugs (If the width of the bushing is short and the plug gauge can pass through, this step can be removed). After that, the right-hand draws out and rotates the plug gauge, lets another end into the hole of the bushing, plugs, and draws out. Then, the left-hand makes the bushing rotate 180 degrees. The right-hand again aims at the hole of the bushing and plugs. When plugging secondly, if it can plug in, the bushing does not meet the dimensional requirement and put it into the waste box. If not, it will be put the finished goods box. The next part will be measured and this process will be repeated.

For the second updating condition, the motion of rotating the plug gauge is not necessary. The others are the same as the first situation.

For the third and fourth updating situation, the right and left hands do the same motions. Reaches to the box of bushings, grasps one respectively, returns on the plug gauge, plugs in, draws out, and puts the parts to the finished goods box or the waste box.

Step 2 Motion analysis

Based on the above mention, the charts of the motion analysis are drawn (Figure 2-10 and Figure 2-11).

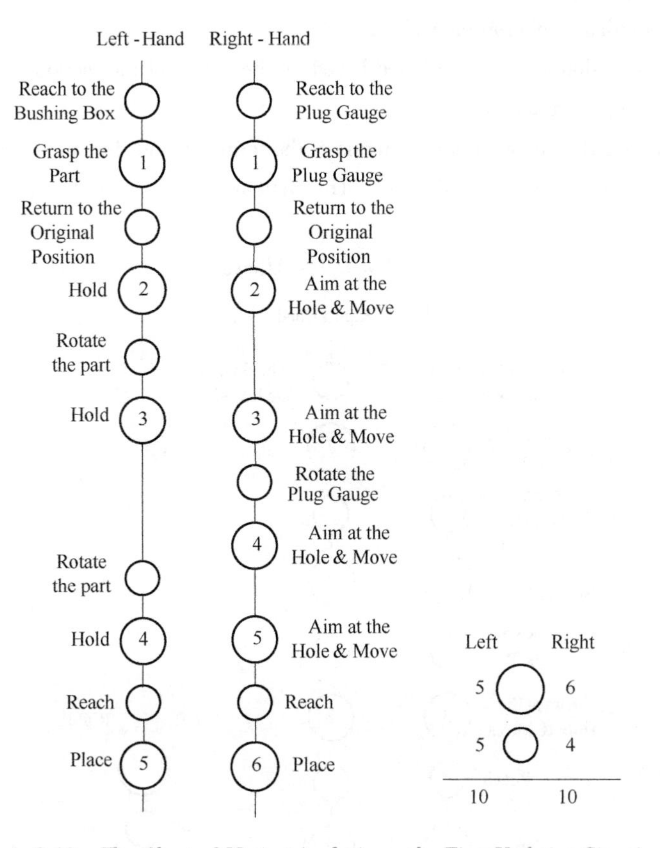

Figure 2-10 The Chart of Motion Analysis on the First Updating Situation

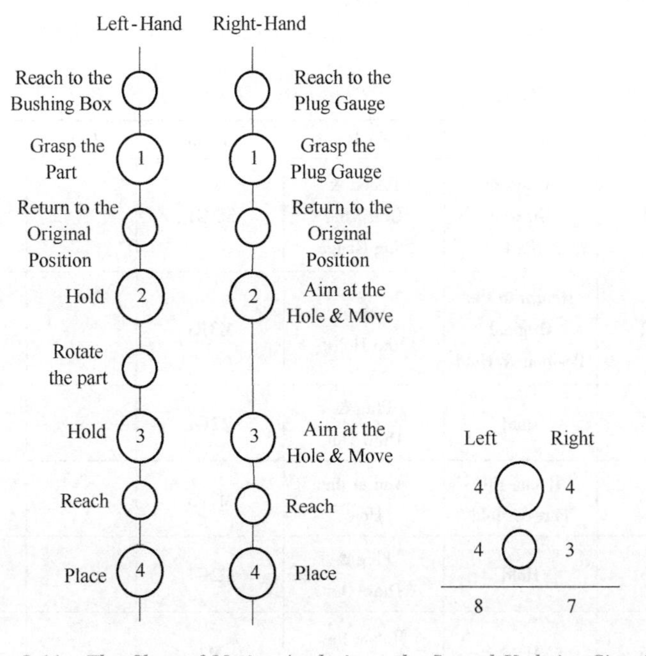

Figure 2-11 The Chart of Motion Analysis on the Second Updating Situation

Step 3 List formulas of motion analysis

The formulas of motion analysis are listed based on the charts of the motion analysis (Table 2-11).

Step 4 Computing operation time

The numbers of MOD are calculated in terms of the formulas and the operation time is known. The computing results are shown in the Table 2-11. The standard time equals the operation time adding allowance.

$$ST = NT \times (1 + Allowance)$$

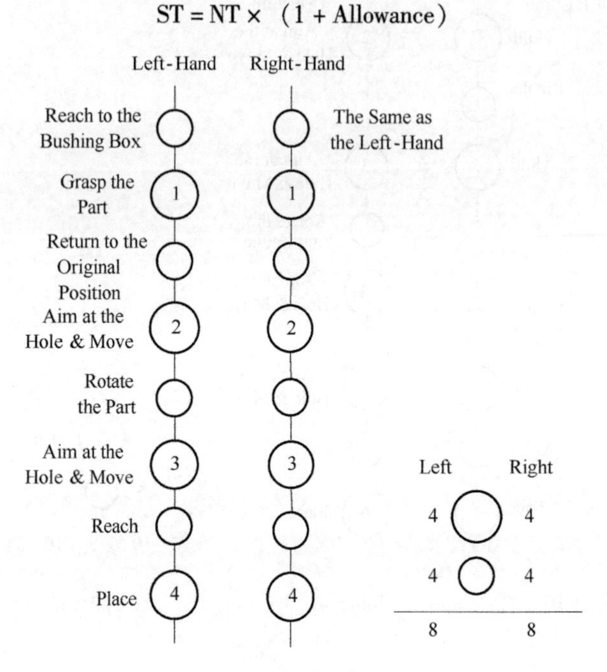

Figure 2-12 The Chart of Motion Analysis on the Last Two Updating Situations

Table 2-11 The Computing Results of MOD

	No.	Left-Hand	Right-Hand	Formula	Time	MOD	Operation Time
The First Updating Situation	1	Reach & Grasp a Part	Reach & Grasp the Plug Gauge	M3G1		4	
	2	Return to the Original Position & Hold	Aim at the Hole	M3P5		8	
	3	Hold	Plug & Draw Out	M2G1	2	6	
	4	Rotate the Part & Hold	Aim at the Hole	M2P5		7	
	5	Hold	Plug & Draw Out	M2G1	2	6	
	6	Hold	Rotate the Plug Gauge	M2		2	

(Contained)

	No.	Left-Hand	Right-Hand	Formula	Time	MOD	Operation Time
The First Updating Situation	7	Hold	Aim at the Hole	M2P5		7	
	8	Hold	Plug & Draw Out	M2G1	1	3	
	9	Rotate the Part & Hold	Aim at the Hole	M2P5		7	
	10	Hold	Plug & Draw Out	M2G1	1	3	
	11	Place the Part	Place the Plug Gauge	M3P0		3	
Total						56	7.2s
The Second Updating Condition	1	Reach & Grasp a Part	Reach & Grasp the Plug Gauge	M3G1		4	
	2	Hold	Aim at the Hole	M3P5		8	
	3	Hold	Plug & Draw Out	M2G1	2	6	
	4	Rotate the Part & Hold	Aim at the Hole	M2P5		7	
	5	Hold	Plug & Draw Out	M2G1	2	6	
	6	Place the Part	Place the Plug Gauge	M3P0		3	
Total						34	4.4s
The Last Two Updating Situations	1	As Right-Hand	Reach & Grasp the Part	M3G1		4	
	2		Aim at the Plug Gauge	M3P5M2		10	
	3		Plug & Draw Out	M2G1	2	6	
	4		Rotate the Part & Aim at the Plug Gauge	M2P5M2		9	
	5		Plug & Draw Out	M2G1	2	6	
	6		Place the Part	M3P0		3	
Total						38	4.9s

Notes

1. When recording the watch readings, note only the necessary digits and omit the decimal point, thus giving as much time as possible to observing the performance of the operator.

句意：当记录秒表读数时，仅记录具体的数字而忽略小数点，因而留出尽可能多的时间观察操作者的表现。

2. Sometimes study researchers use both methods, believing that studies of predominantly long elements are more adapted to snapback readings, while short-cycle studies are better suited to the continuous method.

句意：有时研究者综合使用两种方法，即在主要是长时间单元时采用归零法，而在短周期的研究中则采用连续计时法。

3. The spent time of all the skilled people finishing a piece of basic motion was a constant under the specified conditions.

句意：在指定的条件下，所有熟练人员完成真正基本动作所需要的时间是常量。

4. Therefore, the produced bushings must be measured by the plug gauge.

句意：因此，生产的衬套必须用塞规测量。

Excesses

1. What are the predetermined time systems?

2. Explain what a foreign element is and how foreign elements are handled under the continuous method.

3. Indicate the advantages and disadvantages of the continuous timing method and snapback method respectively.

4. Explain the characteristics of the predetermined time systems.

5. In this Unit, we introduce how many methods of the predetermined time systems. Give a brief description respectively.

6. Give a case to explain how to use MOD method.

Unit 4 Work Sampling

Concept of Work Sampling

Work sampling is also called the instant observation method. It gets the required results through observing instantly, systematic analysis and dealing with the observation data to the operator during a longer period.

The difference between work sampling and stopwatch time study can be indicated with an example of observing the ratio of working time and vacancy time for an operator within one hour.

Stopwatch time study

In one hour, the direct time observation was carried out with one minute as a time unit and observation results were recorded as shown in Figure 2-13. The blank white grids represent working time, the oblique grids show vacancy time. Based on the Figure 2-13, within 60 minutes, there are 18 minutes vacancy and 42 minutes working.

Work ratio = work time/total observation time = 42/60 = 70%

Vacancy ratio = vacancy time/total observation time =18/60 =30%

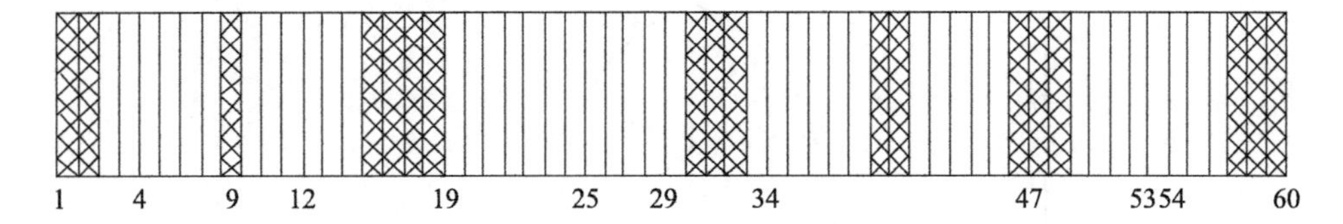

Figure 2-13 The Results of Observations Record

Work sampling

Take one hour divide into 60 grids, then, 10 numbers within 1 to 60 were selected randomly as the observations. Supposed the following numbers were selected:

34, 54, 4, 47, 53, 29, 12, 9, 19, 25

They were arrayed from small to big:

4, 9, 12, 19, 25, 29, 34, 47, 53, 54

That means that the observation was done at 4th minute, 9th minute, 12th minute, etc. within 60 minutes. Totally 10 times were observed and the results were recorded. Among them 3 times are vacancies and 7 times are working.

Work ratio = work time/total observation time = 7/10 = 70%

Vacancy ratio = vacancy time/total observation time =3/10 =30%

It can be seen that the observed results are the same for both stopwatch time study and work sampling.

Applications of Work Sampling

Work sampling can be used in the following two situations:

Work improvement. With work sampling, the work ratio and vacancy ratio of the operator or machine could be observed and calculated. To study the vacancy portions, the problems could be found, and the improvements could be done.

Setup time standards. With work sampling, the time standard could be set up by the following formulations:

Standard Time = (Total observation time × work ratio × average performance index) /Total observation time + allowance

Average performance index (%) = the normal time spent by the product/the practical time spent by the product

Accuracy and Observations

Based on the theory of probability, the phenomena disposed by work sampling submit to the normal distribution approximately[1] as shown in Figure 2-14.

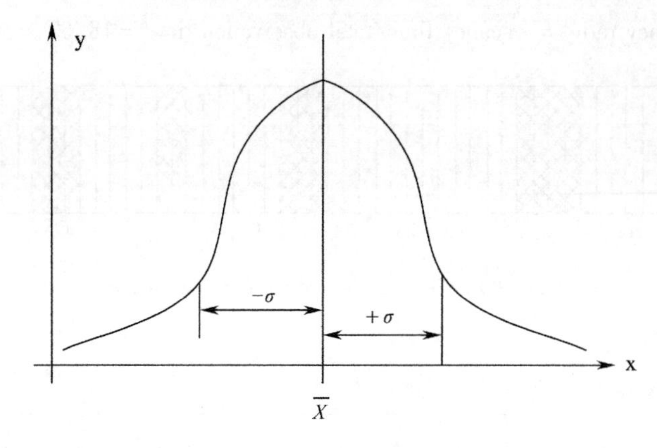

Figure 2-14　The Normal Distribution Chart

When let the average number $\overline{X}$ as a center and take the twice of standard deviation σ, 95 percent confidence could be obtained. That is to say, 95 samples are approximately similar to the status of the population among 100 samples. In other words, there are 95 percent predetermined sampling data located in the scope of $\overline{X} \pm 2\sigma$, just only 5 percent data are beyond this scope.

Based on the principles of statistics, the standard deviation of the binomial distribution equals the following equation under the certain conditions:

$$\sigma = \sqrt{\frac{P(1-P)}{n}}$$

Accuracy means the allowed error. Sampling accuracy includes the absolute accuracy E and the relative accuracy S. When the confidence value is set up to 95 percent, the absolute accuracy E can be calculated by the following equation.

$$E = 2\sigma = 2\sqrt{\frac{P(1-P)}{n}} \tag{2-5}$$

P——Observed probability;

n——Observations.

The relative accuracys equals to:

$$S = \frac{E}{P} = 2\sqrt{\frac{1-P}{nP}} \tag{2-6}$$

The different standards of the absolute accuracy are used based on the different sampling purposes in work sampling as shown in the Table 2-12.

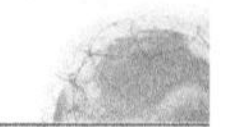

Table 2-12 Sampling Purposes and the Standards of the Absolute Accuracy

The Sampling Purposes	The Standards of the Absolute Accuracy
Survey of Breakdown	± (3.6 - 4.5)%
Work Improvement	± (2.4 - 3.5)%
Decision of the Time Ratios	± (1.2 - 1.4)%
Setting Up the Standard Time	± (1.6 - 2.4)%

The standards of the relative accuracy are selected between ± (5 - 10)%. In general, if the confidence value is set up to 95%, the relative accuracy equals to ±5%.

The observations are calculated based on the confidence and accuracy values. When the confidence value equals to 95%, the observations could be obtained by the formulas (2-5) or (2-6).

Using the absolute accuracy E,

$$n = \frac{4P(1-P)}{E^2} \tag{2-7}$$

Using the relative accuracy S,

$$n = \frac{4(1-P)}{S^2 P} \tag{2-8}$$

In the formulas (2-7) and (2-8), each has two unknown parameters P and n. We could not get the solutions with one equation. First, 100 observations are carried out to try to get P. For example, through 100 observations, if the work ratio of the equipment is 75% and the absolute accuracy sets up to ±3%, the observed times equal to:

$$n = \frac{4P(1-P)}{E^2} = \frac{4 \times 0.75(1-0.75)}{0.03^2} = 334\text{observations}$$

The observations calculated by the formulas (2-7) and (2-8) are the required samples. If there are X observers or X machines, one observation can get X samples. Therefore, the actual observations $K = n/X$.

For example, there is a work group, which has 10 workers, the confidence sets up to 95%, the relative accuracy is 5%, the work ratio is 70%, 20 observations are planned every day, ask for how many days are required to do observation.

$$n = \frac{4(1-P)}{S^2 P} = \frac{4 \times (1-0.7)}{0.05^2 \times 0.7} = 686\text{observations}$$

The actual observations $K = n/X = 686/10 = 68.6 \approx 69$observations

The observation days $= 69/20 \approx 4$ days

The Method of Work Sampling

The method of work sampling includes nine steps.

Step 1 Planning the survey purposes and scope

Because there are different survey purposes, there are different survey ways and observed times as well as different classifications of the survey items.[2] For example, if you want to survey the machine operations, you should make clear that what machine or machines should be surveyed such as one machine, several machines, machines within one shop floor, or machines in the factory.

Step 2 Classification of the survey items

Based on the survey purposes, the activities of survey items can be classified. If you want to survey the machine work ratio, the observation items may be "working", "shutdown", and "idleness" as shown in Figure 2-15. If you want to know the reasons of shutdown and idleness, the classification should include them.

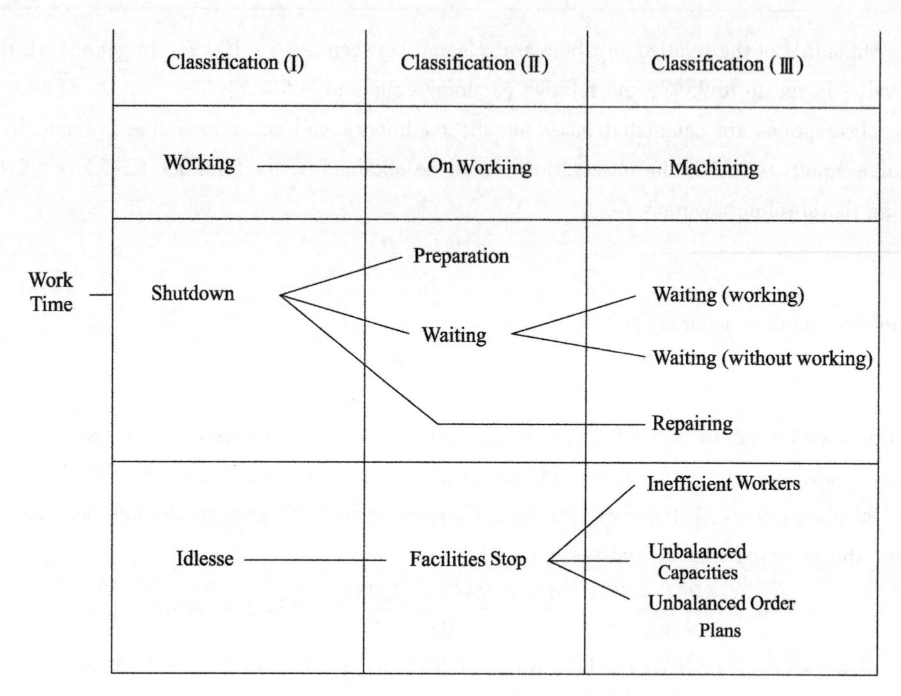

Figure 2-15　The Classifications of Survey Items

The classification of sampling items is the basis of work sampling table design. It is also a guarantee of making the sampling results to meet the sampling purposes.

Step 3 Decision of observed methods

Before observation, the distribution layout of the observed machines or operators and the routings of the itinerant observation should be drawn.[3] The observation position should be indicated.

Step 4 Design of the survey table

The contents and formats of the survey table depend on the survey purposes and requirements. If you use the Table 2-13, you just know the work ratios of the machine and operator but do not know the reasons of idleness.

Table 2-13　Analysis of Idleness Time

Classification		Operations	Idleness	Totals	Ratio
Machine	1	32	18	50	64%
	2	38	12	50	76%
	3	24	26	50	48%

(Contained)

Classification		Operations	Idleness	Totals	Ratio
Operator	1	27	23	50	54%
	2	22	28	50	44%
	3	33	17	50	66%

Step 5 Explanation of survey purposes

In order to ensure the success of the work sampling, the operators should be told about the survey purposes, significances, and the operational methods to remove the operator's worries and misunderstandings. At the same time, the operators are required to work in normal way.

Step 6 Pre-observation and decision of the observations

When using the formulas (2-7) or (2-8), the observed probability P should be known, and then n can be calculated. P is determined by pre-observation (to see the last section).

Step 7 Formal observations

When the observations are done, the observed timetable every day must be determined at first. The decision of the observed timetable should ensure that the observation is random. It is the theoretical basis of work sampling. If the unsuitable observed timetable is selected, the observation deviation should be produced. There are many different methods to determine the observed timetable. The randomly sampling and the layered sampling are the two methods used frequently.

The Randomly Sampling

[**Example 1**]

The work sampling is done on the shop floor of the factory. 5 days are observed and 20 observations are carried out every day. The working time on the job is a full 8-hour day (one hour rest from 12:00 to 13:00). Ask for the observed timetable every day.

1. Arranging random numbers in two-digital figures. The simplest method is that take a piece of the yellow paper representing the first digital figures from 0 to 9, take a piece of the red paper representing the decimal digital figures from 0 to 9. Select randomly one piece of the paper from each color, record the figure, return the pieces of the paper, and select again. Repeat this process and a random figures list is obtained. For this sample, there are 15 selections, the random figures list shows as follows:

21, 94, 62, 35, 06, 64, 96, 40, 85, 77, 88, 63, 52, 27, 75

2. Remain the figures that are less than 50 and the residues subtract 50. We get:

21, 44, 12, 35, 06, 14, 46, 40, 35, 27, 38, 13, 02, 27, 25

3. Remove the figures that more than 30. We get:

21, 12, 06, 14, 27, 13, 02, 27, 25

4. Determine the observed timetable on the first day. At first, we want to decide the first timetable of observation on the first day. The first number is 21. Because of the working time is at 8:00, the first timetable of observation is at 8:21. Then we want to decide the intervals of observation.

The interval = (the working time − the first timetable of observation) /the observations = $(480-21)/20 \approx 23$

In this way, we can get the timetable of observation on the second day

8:21 + 0:23 = 8:44

The timetable of observation on the third day

8:44 + 0:23 = 9:07

5. Decide the timetable of observation on the second day. At first, we want to decide the first timetable of observation on the second day. The second number is 12. Because of the working time is at 8:00, the first timetable of observation is at 8:12. Based on the interval of 23, we can get the second timetable of observation is 8:35, the third timetable of observation is 8:58, and so on.

6. Decide the timetables of observation on the third day through the fifth day. The calculation method is the same. The calculated results are shown in the Table 2-14.

Table 2-14 Observation Timetables

Observation Days		1	2	3	4	5
Random Numbers		21	12	06	14	27
The First Timetable of Observation		8:21	8:12	8:06	8:14	8:27
Interval		23	23	23	23	23
Observed Times	1	8:21	8:12	8:06	8:14	8:27
	2	44	35	39	37	50
	3	9:07	58	52	9:00	9:13
	4	30	9:21	9:15	23	36
	5	53	44	38	46	59
	6	10:16	10:07	10:01	10:09	10:22
	7	39	30	24	32	45
	8	11:02	53	47	55	11:08
	9	25	11:16	11:10	11:18	31
	10	48	39	33	41	54
	11	13:21	13:12	13:06	13:14	13:27
	12	44	35	29	37	50
	13	14:07	58	52	14:00	14:13
	14	30	14:21	14:15	23	36
	15	53	44	38	46	59
	16	15:16	15:07	15:01	15:09	15:22
	17	39	30	24	32	45
	18	16:02	53	47	55	16:08
	19	25	16:16	16:10	16:18	31
	20	48	39	33	41	54

The Layered Sampling

Unlike the randomly sampling, the layered sampling determines the observed timetable based on the classifications of working time.

[**Example 2**]

Table 2-15 is the working schedule of one shift on the shop floor at a factory. Try to decide the

observations and timetables.

Table 2-15 The Working Schedule

Working Duration	Working Task
8:00 - 8:30	Preparation and machine setup
8:30 - 11:45	Working
11:45 - 12:00	Arrangement
13:00 - 13:15	Preparation
13:15 - 16:30	Working
16:30 - 17:00	Ending, clearing, and cleaning

In this case, we want to decide the observations and timetables in different time durations. It is called the layered sampling. [4]

Suppose the total observed times on one day are 60, working time is 8 hours (480 minutes). The observations can be calculated by the Table 2-16.

Table 2-16 The Observations

Working Duration	Observations
8:00 - 8:30	(30/480) ×60 =4
8:30 - 11:45	(195/480) ×60 =24
11:45 - 12:00	(15/480) ×60 =2
13:00 - 13:15	(15/480) ×60 =2
13:15 - 16:30	(195/480) ×60 =24
16:30 - 17:00	(30/480) ×60 =4

The above two methods mentioned above could be selected in terms of actual conditions.

The observers conduct observations based on the observed timetable and the survey items. When observing, the observed data are recorded on the survey table accurately.

Step 8 Data disposal

(1) Removing the abnormal data.

After the observed data are recorded, the management chart is drawn based on the recorded data and the management limitations are determined. The data that are beyond the limitations should be removed. The management limitations are calculated by the equation 2-9.

$$\text{The management limitations} = P \pm \sqrt{\frac{P\ (1-P)}{n}} \tag{2-9}$$

P—the average of the observed event probability;

n—the observations every day.

Let us give an example to explain how to remove the abnormal data from the observed data.

Table 2-17 shows an observed result of working sampling on the production line. Try to remove the abnormal data.

Table 2-17 An Observation Result of Working Sampling

Observed Shifts	Observations Every Shift	Working Times	Working Ratios
1	160	129	80.63
2	160	142	88.75
3	160	124	77.50
4	160	125	78.13
5	160	119	74.38
6	160	120	75.00
Total	960	759	79.06

The management limitations =

$$P \pm \sqrt{\frac{P(1-P)}{n}} = 0.7906 \pm \sqrt{\frac{0.7906(1-0.7906)}{160}} = 0.7906 \pm 0.0966$$

The upper limitation = 0.7906 + 0.0966 = 0.8872

The low limitation = 0.7906 - 0.0966 = 0.6940

The management chart is drawn in Figure 2-16.

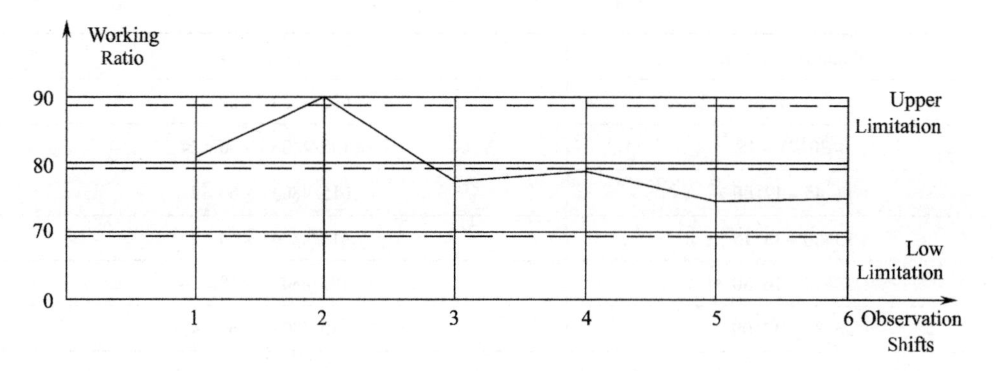

Figure 2-16 Management Chart

We find from the Figure 2-16 that the working ratio of the second shift is 88.75% beyond the management upper limitation 88.72%. Therefore, this datum acted as an abnormal datum should be removed.

(2) Verifying observations and accuracy.

When the abnormal data being removed, the new average probability of the observed events should be calculated again. The confidence of the average probability should be determined by the original confidence and accuracy values.

After the abnormal data being removed, we want to determine if the left observations satisfy the total observations. If not, new observations should be done.

After the abnormal data being removed, the absolute and relative accuracy values in terms of the new average probability should be calculated again and are compared with the original accuracy requirements. If these values do not satisfy the requirements, new observations should be done.

In the last sample, after the data of the second shift being removed, the new average probability is calculated:

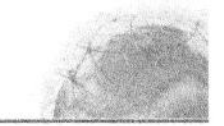

$$\overline{P} = \frac{working\ time}{total\ obaervation\ time} = \frac{129+124+125+119+120}{160\times 5}\times 100\% = 77.13\%$$

After the abnormal data being removed, the left observations are 800, the original specified observations are 400. Therefore, new observation is unnecessary.

The new absolute accuracy value:

$$E = 2\sqrt{\frac{P\ (1-P)}{n}} = 2\sqrt{\frac{0.7713\times\ (1-0.7713)}{160\times 5}} = \pm 0.0279$$

The new relative accuracy value:

$$S = 2\sqrt{\frac{(1-P)}{n}} = 2\sqrt{\frac{1-0.7713}{160\times 5}} = \pm 0.0388$$

Suppose the originally absolute accuracy value is 3%, the originally relative accuracy value is 5%. The new accuracy values are satisfied.

Step 9 Making conclusions

Based on the above eight steps, we can know the observed results and give the conclusion, such as if the working ratio is suitable, if working load is full, if the workers are enough, and so on. After the conclusion being made, the reasons that result in the problems should be analyzed, the future solution should be put out to meet the objectives of both playing the labor and equipment potentials and improving profits.

Notes

1. Based on the theory of probability, the phenomena disposed by work sampling submit to the normal distribution approximately.

句意：根据概率论，用工作抽样法处理的现象接近于正态分布。

2. Because there are different survey purposes, there are different survey ways and observed times as well as different classifications of the survey items.

句意：调查目的不同，则调查项目分类、调查方法和观测次数均不相同。

3. Before observation, the distribution layout of the observed machines or operators and the routings of the itinerant observation should be drawn.

句意：在观测前，要绘制被观测设备和操作者分布平面图和巡回观测的路线图。

4. In this case, we want to decide the observations and timetables in different time durations. It is called the layered sampling.

句意：根据不同的时间间隔决定观测的次数和观测时间，就称为分层抽样。

Exercises

1. What steps are taken for work sampling?

2. The work sampling is done on the shop floor of the factory. 5 days are observed and 20 observations are carried out every day. The working time on the job is a full 8-hour day (one hour rest from 12:00 to 13:00). Ask for the observed timetables every day with the randomly sampling method.

3. A work measurement researcher in the Dongfang Company took 12 observations of a high-per-

formance job. His performance rated each cycle and then computed the mean normal time for each element. The element with the greatest dispersion had a mean of 0. 30 minutes and a standard deviation of 0. 003 minutes. If it is desirable to have sampled data within ±5 percent of the time data, how many observations should this time study researcher take of this operation?

4. In the Dongfang Company, the work measurement researcher took a detailed time study of the making shell molds. The third element of this study has the greatest variation in time. After study nine cycles, the research computed the mean and the standard deviation of this element, with the following results: $X = 0.42$, $\sigma = 0.08$. if the researcher wanted to be 95 percent confident that the mean time of the sample was within ±5 percent of the mean of the population, how many total observations should have been taken? Within what percent of the average of the total population is x at the 97 percent confidence level, under the measured observations?

5. How to remove the abnormal data when the observed data are disposed.

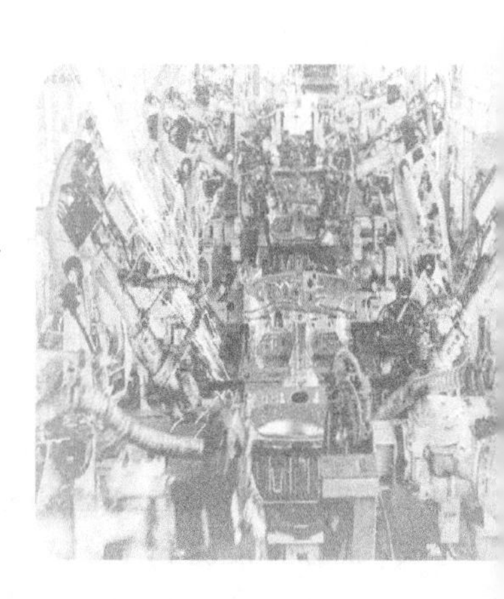

Chapter 3 Manufacturing Systems

Unit 1 Introduction to Manufacturing Systems

In this chapter, we consider how automation and material handling technologies are synthesized to create manufacturing systems. We define a manufacturing system to be a collection of integrated equipment and human resources, whose function is to perform one or more processing and/or assembly operations on a starting raw material, part, or set of parts.[1] The integrated equipment includes production machines and tools, material handling and work positioning devices, and computer systems. Human resources are required either full time or periodically to keep the system running. The manufacturing system is where the value-added work is accomplished on the part or product. The position of the manufacturing system in the production system as shown in Figure 3-1. Examples of manufacturing systems include:

- One worker tending one machine, which operates on semi-automatic cycle;
- A cluster of semi-automated assembly machine, attended by one worker;
- A full automated assembly machine, periodically attended by a human worker;
- A group of automated machines working on automatic cycles to produce a family of similar parts;
- A team of workers performing assembly operations on a production line.

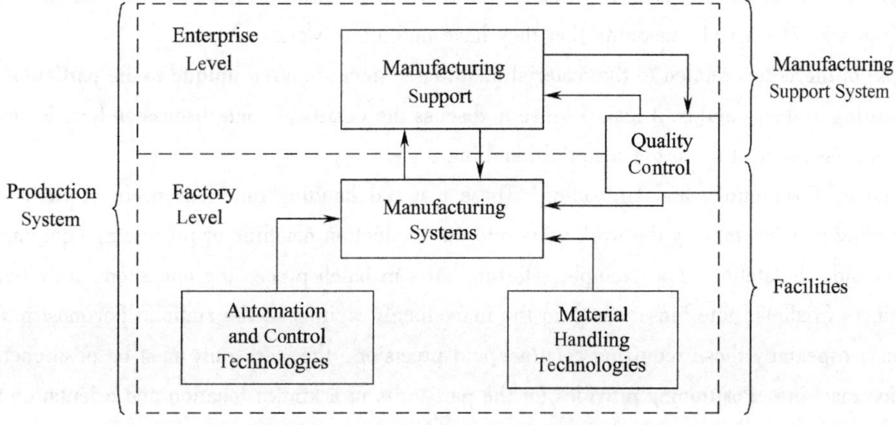

Figure 3-1 The Position of the Manufacturing System in the Production System

Components of a Manufacturing System

A manufacturing system consists of several components. In a given system, these components usually include: ① production machines plus tools, fixtures, and other related hardware; ② material handling system; ③ computer systems to coordinate and/or control the above components; ④ human workers.

Production Machines

In virtually all modern manufacturing systems, most of the actual processing or assembly work is accomplished by machines or with the aid of tools. The machines can be classified as ① manually operated, ② semi-automated, or ③ fully automated. *Manually operated machines* are directed or supervised by a human worker. The machine provides the power for the operation and the worker provides the control. Conventional machine tools (e. g., lathes, milling machines, drill presses) fit into this category. The worker must be at the machine continuously.

In manufacturing systems, we use the term workstation to refer to a location in the factory where some well-defined task or operation is accomplished by an automated machine, a worker-and-machine combination, or a worker using hand tools/or portable powered tools. In the last, there is no definable production machine at the location. Many assembly tasks are in this category. A given manufacturing system may consist of one or more workstations. A system with multiple stations is called a production line, or assembly line, or machine cell, or other name, depending on its configuration and function.

Material Transport Systems

In most processing and assembly operations performed on discrete parts and products, the following ancillary functions must be provided: ① loading and unloading work units and ② positioning the work units at each station. In manufacturing systems composed of multiple workstations, a means of ③ transporting work units between stations is also required. These functions are accomplished by the material handling system. In many cases, the units are moved by the workers themselves, but more often some form of mechanized or automated material transport system is used to reduce human effort. Most material handling systems used in production also provide ④ a temporary storage function. The purpose of storage in these systems is usually to make sure that work is always present for the stations, that is, the stations are not starved (meaning that they have nothing to work on).

Some of the issues related to the material handling system are often unique to the particular type of manufacturing system, and so it makes sense to discuss the details of each discussion here is concerned with general issues relating to the material handling system.

Loading, Positioning, and Unloading. These material handing functions occur at each workstation. Loading involves moving the work units into the production machine or processing equipment from a source inside the station. For example, starting parts in batch processing operations are often stored in containers (pallets, tote bins, etc.) in the immediately vicinity of the station. For most processing operations, especially those requiring accuracy and precision, the work unit must be positioned in the production machine. Positioning provides for the part to be in a known location and orientation relative to the workhead or tooling that performs the operation.

Positioning in the production equipment is often accomplished using a workholder. A workholder is a device that accurately locates, orients, and clamps the part for the operation and resists any forces that may occur during processing. Common workholders include jigs, fixtures, and chucks. When the production operation has been completed, the work unit must be unloaded, that is, removed from the production machine and either placed in a container at the workstation or prepared for transport to the next workstation in the processing sequence. "Prepared for transport" may consist of simply loading the part onto a conveyor leading to the next station.

When the production machine is manually operated or semi-automatic, loading, positioning, and unloading are performed by the worker either by hand or with the aid of a hoist. A mechanized device such as an industrial robot, parts feeder, coil feeder (in sheet metal stamping), or automatic pallet changer is used to accomplish these material handling functions.

Work Transport Between Stations. In the context of manufacturing system, work transport means moving parts between workstations in a multi-station system. The transport function can be accomplished manually or by the most appropriate material transport equipment.

In some manufacturing systems, work units are passed from station to station by hand. Manual work transport can be accomplished by moving the units one at a time or in batches. Moving parts in batches is generally more efficient, according to the unit load principle. Manual work transport is limited to cases in which the parts are small and light, so that the manual labor is ergonomically acceptable. When the load to be moved exceeds certain weight standards, powered hoists and similar lift equipment are used. Manufacturing systems that utilize manual work transport include manual assembly lines and group technology machine cells.

Various types of mechanized and automated material handling equipment are widely used to transport work units in manufacturing systems. We distinguish two general categories of work transport, according to the type of routing between stations: ① variable routing and ② fixed routing. In variable routing transport is associated with job shop production and many batch production operations. Manufacturing systems that use variable routing include group technology machine cells and flexible manufacturing systems. In fixed touting, the work units always flow through the same sequence of stations. This means that the work units are identical or similar enough that the processing sequence is identical. Fixed routing transport is used on production lines. The difference between variable and fixed routing is portrayed in Figure 3-2.

Computer Control System

In today's automated manufacturing systems, a computer is required to control the automated and semi-automated equipment and to participate in the overall coordination and management of the manufacturing system. Even in manually driven manufacturing systems, such as a completely manual assembly line, a computer system is useful to support production. Typical system functions include the following:

Communicate instructions to workers. In manually operated workstations that perform different work units, processing or assembly instructions for the specific work units must be communicated to the operator.

Download part programs to computer-controlled machines (e. g., CNC machine tools).

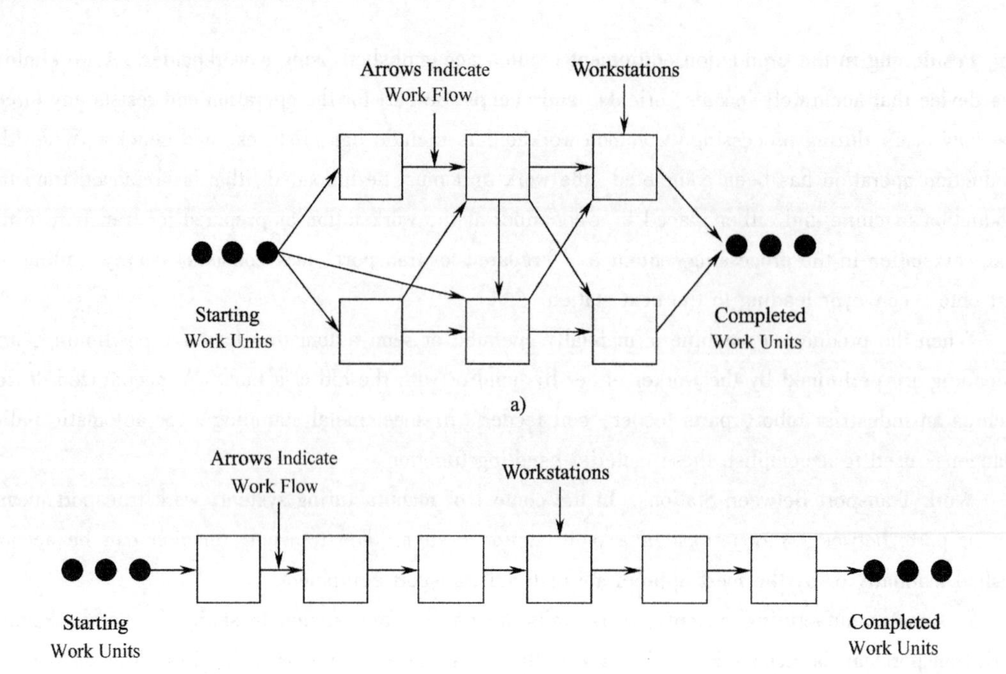

Figure 3-2 Types of Routing in Multiple Stations Manufacturing Systems

a) Variable Routing b) Fixed Routing

Material handling system control. This function is concerned with controlling the material handling system and coordinating its activities with those of the workstations.

Schedule production. Certain production scheduling functions are accomplished at the site of the manufacturing system.

Failure diagnosis. This involves diagnosing equipment malfunctions, preparing preventive maintenance schedules, and maintaining spare parts inventory.

Safety monitoring. This function ensures that the system does not operate in an unsafe condition. The goal of safety monitoring is to protect both the human workers manning the system and the equipment comprising the system.

Quality control. The purpose of this control function is to defect and possibly reject defective work units produced by the system.

Operations management. Managing the overall operations of the manufacturing system, either directly (by supervisory computer control) or indirectly (by preparing the necessary reports for management personnel).

Human Resources

In many manufacturing systems, human perform some or all of the value-added work that is accomplished on the parts or products. In these cases, the human workers are referred to as direct labor. Though their physical labor, they directly add to the value of the work unit by performing manual work on it or by controlling the machines that perform the work. In manufacturing systems that are fully automated, direct labor is still needed to perform such activities as loading and unloading parts to and from

the system, changing tools, resharpening tools, and similar functions. Human workers are also needed for automated manufacturing systems to manage or support the system as computer programmers, computer operators, and part programmers for CNC machine tools, maintenance and repair personnel, and similar indirect labor tasks. In automated systems, the distinction between direct and indirect labor is not always precise.

Classification of Manufacturing Systems

First of all, manufacturing systems are distinguished by the types of operations they perform. At the highest level, the distinction is between ① processing operations on individual work units and ② assembly operations to combine individual parts into assembled entities. Beyond this distinction, there are the technologies of the individual processing and assembly operations.

Additional parameters of the product that play a role in determining the design of the manufacturing system include: type of material processed, size and weight of the part or product, and part geometry. For example, machined parts can be classified according to part geometry as rotational or non-rotational. Rotational parts are cylindrical or disk-shaped and require turning and related rotational operations. Non-rotational (also called prismatic) parts are rectangular or cube-like and require milling and related machining operations to shape them. Manufacturing systems that perform machining operations must be distinguished according to whether they make rotational or non-rotational parts. The distinction is important not only because of differences in the machining processes and machine tools required, but also because the material handling system must be engineered differently for the two cases.

Type 1 Manufacturing Systems: Single Stations

Applications of single workstations are widespread. The typical case is a worker-machine cell. Our classification scheme distinguishes two categories: ① manned workstations in which a worker must be in attendance either continuously or for a portion of each work cycle, and ② automated stations, in which periodic attention is required less frequently than every cycle. In either case, these systems are used for processing as well as for assembly operations, and their applications include single model, batch model, and mixed model production.

Reasons for the popularity of the single model workstation include: ① it is the easiest and least expensive manufacturing method to implement, especially the manned version; ② it is the most adaptable, adjustable, and flexible manufacturing system; and ③ a manned single workstation can be converted to an automated station if demand for the parts or products made in the station justifies this conversion.

Type 2 Manufacturing System: Multi-Station Cells

A multiple station system with variable routing is a group of workstations organized to achieve some special purpose. It is typically intended for production quantities in the medium range (annual production = $10^2 \sim 10^4$ parts or products), although its applications sometimes extend beyond these boundaries. The special purpose may be any of the following:

- Production of a family of parts having similar processing operations;
- Assembly of a family of products having similar assembly operations;

• Production of the complete set of components used in the assembly of one unit of final product. By producing all of the parts in one product, rather than batch production of the parts, work-in-process inventory is reduced.

The multi-station system with variable routing is applicable to either processing or assembly operations. It also indicates that the applications usually involve a certain degree of part or product variety, which means differences in operations and sequences of operations that must be performed. The machine groups must possess flexibility to cope with this variety.

The machines in the group may be manually operated, semi-automatic, or fully automated. In our classification scheme, manually operated machine groups are type 2. These groups are often called machine cells, and the use of these cells in a factory is called cellular manufacturing. Cellular manufacturing and its companion topic, group technology, are discussed later. When the machines in the group are fully automated, with automated material handling between workstations, it is also classified as type 2. If an automated machine group is flexible, it is referred to as a flexible manufacturing system or flexible manufacturing cell.

Type 3 Manufacturing System: Production Lines

A multi-station manufacturing system with fixed routing is a production line. A production line consists of a series of workstations laid out so that the part or product moves from one station to the next, a portion of the total work is performed on it at each station. Production lines are generally associated with mass production ($10^4 \sim 10^6$ parts or products per year). Conditions that favor the use of a production line are:

• The quantity of parts or products to be made is very high (up to millions of units);

• The work units are identical or very similar; (Thus they require the same or similar operations to be performed in the same sequence.)

• The total work can be divided into separate tasks of approximately equal duration that can be assigned to individual workstations.

The production rate of the line is determined by its slowest station. Workstations whose pace is faster than the slowest must ultimately wait for that bottleneck station. Transfer of work units from one station to the next is usually accomplished by a conveyor or other mechanical transport system, although in some cases the work is simply pushed between stations by hand.

Production lines are used for either processing or assembly operations. It is unusual for both types of operation to be accomplished on the same line. Production lines are either manually operated or automated.

Tendency of Manufacturing Systems

The modern manufacturing system—a complex arrangement that engineers, manufactures, and markets products for a diverse populace whose wants and needs are subject to frequent change—bears little similarity to its early predecessor. This transformation of the manufacturing function into the complex technical, social, and economic organization it is today has been made possible by the discovery and improvement of many methods and processes. Some of these have revolutionized the technologies employed in the transformation of materials; some have made possible the creation of sophisticated organizations for the design, production, and marketing of products and services; and some have made it possible to create new and unique materials that have expanded design alternatives and formed the basis for entirely

new products, processes, and industries.

These developments have their origins in many activities, some internal to manufacturing and some well removed from it. It is not an exaggeration, however, to claim that some of the most profound changes that have affected the manufacturing enterprise can be traced to major developments that resulted from the complex interactions of society, technology, and the economy.

Three of the most important of these are widespread and inexpensive transportation systems, communication systems that provide real-time interaction between people in almost any part of the world, and computers that assist in the design, control, and analysis of complex activities.[2] Given the ability to ship most finished goods quickly and inexpensively almost anyplace in the world, companies no longer must locate their factories in or near the markets they serve. Neither do manufacturers require an extensive national industrial infrastructure to support local or regional factories with low levels of manufacturing integration.

Modern communication technology has made it possible to manage manufacturing operations located around the world. A highly dispersed network of facilities and suppliers can now be created irrespective of their location. For example, the local availability of raw materials is not a determinant or predictor of successful manufacturing capabilities. Information concerning technological developments flows virtually unimpeded across national boundaries. Finally, the availability of modern computer technology has made it possible to analyze, design, and control the complex systems that characterize modern manufacturing operations.

These three technical developments have had a profound effect on both the organization of the manufacturing enterprise and the strategies that manufacturing enterprises follow to survive. A consequence of these new capabilities is that manufacturers can now establish an effective presence in almost any market that will accept their products.

Notes

1. We define a manufacturing system to be a collection of integrated equipment and human resources, whose function is to perform one or more processing and/or assembly operations on a starting raw material, part, or set of parts.

句意：我们把一个制造系统定义为设备与人力资源的一个集合，其功能是执行一个或多个始于原材料的加工过程和/或始于零件或部件的装配过程。

2. Three of the most important of these are widespread and inexpensive transportation systems, communication systems that provide real-time interaction between people in almost any part of the world, and computers that assist in the design, control, and analysis of complex activities.

句意：在社会、技术和经济复杂交互活动中的主要发展包括三个方面：分布广泛且经济的运输系统，给几乎世界任何地方的人们提供及时交互的通信系统，以及计算机在辅助设计、控制和综合活动分析方面的应用。

Exercises

1. What is the manufacturing system? And how do you define it in your own opinion?

2. How many components does a manufacturing system have? And which one is more important in a manufacturing system? Why?

3. How many types does a manufacturing system have? And what is the difference between them?

4. Recently what makes a manufacturing system change so much? Why?

5. What is the most important for modern manufacturers who can establish an effective market that will accept their product in almost any place in the world?

Unit 2 Advanced Manufacturing Systems

Group Technology

Group Technology (GT) is a manufacturing philosophy in which similar parts are identifies and grouped together to take advantage of their similarities in design and production. Similar parts are arranged into part families. Where each part family possesses similar design and/or manufacturing characteristics. For example, in a plant there are over 10,000 different kinds of parts to be produced. These parts may be able to group into 30 – 40 distinct families. It is reasonable to believe that the processing of each member of a given family is similar and this should result in manufacturing efficiencies. The efficiencies are generally achieved by arranging the production equipment into machine groups, or cells, to facilitate in flow. Grouping the production equipment into machine cells, where each cell specialization is an example of mixed model production.

Two Major Tasks

There are two major tasks that a company must undertake when it implements group technology. These two tasks represent significant obstacles to the application of GT. ① Identifying the part families. If the plant makes 10,000 different parts, reviewing all of the part drawings and grouping the parts into families is a substantial task that consumes a significant amount of time. ② Rearranging production machines into machine cells. It is time consuming and costly to plan and accomplish this rearrangement, and the machines do not work during the changeover.

Benefits

Group technology offers substantial benefits to companies that have the perseverance to implement it. The benefits include: ① GT promotes standardization of tooling, featuring, and setups. ② Material handling is reduced because parts are moved within a machine cell rather than within the entire factory. ③ Process planning and production scheduling are simplified. ④ Setup times are reduced, resulting in lower manufacturing lead times. ⑤Work-in-process is reduced. ⑥ Worker satisfaction usually improves when workers collaborate in a GT cell. ⑦ Higher quality work is accomplished using group technology.

Flexible Manufacturing Systems

A Flexible Manufacturing System (FMS) is a system of numerical control (NC) machine tools which has automated material handling and central so that it has an automatic tool handling capacity.

Because of its automatic tool handling capability and computer control, such a system can be continually reconfigured to manufacture a wide variety of parts. This is why it is called a flexible manufacturing system.

Content of FMS

A FMS typically encompasses: ① Process equipment, e. g. , machine tools, assembly stations, and robots; ② Material handling equipment, e. g. , robots, conveyors, and AGVs (Automated Guided Vehicles); ③ A communication system; ④ A computer control system.

Future FMS will contain many manufacturing cells, each cell consisting of robot serving several computers NC (CNC) machine tools or other stand-alone systems such as an inspection machine, a welder, an Electron Discharge Machining (EDM) machine, etc.[1] The manufacturing cells will be located along a central transfer system, such as a conveyor, on which a variety of different workpieces and parts are moving. The production of each part will require processing through a different combination of manufacturing cells. In many cases more than one cell can perform a given processing step. When a specific workpiece approaches the required cell on the conveyor, the corresponding robot will pick it up and lead it onto a CNC machine in the cell. After processing in the cell, the robot will return the semi-finished or finished part to the conveyor. A semi-finished part will move on the conveyor until it approaches a subsequent cell where its processing can be continued. The corresponding robot will pick it up and load it onto a machine tool. This sequence will be repeated along the conveyor, until, at the route, there will be only finished parts moving, and then they could be routed to an automatic inspection station and subsequently unloaded from the FMS. The coordination among the manufacturing cells and the control of the part's flow on the conveyor will be accomplished under the supervision of the central computer.

Advantages of FMS

The advantages of FMS include the following: ① Increased productivity; ② Shorter preparation time for new products; ③ Reduction of inventory parts in the plant; ④ Saving of labor cost; ⑤ Improves product quality; ⑥ Attracting skilled people to manufacturing (since factory work is regarded as boring and dirty); ⑦ Improved operator's safety.

Additional economic savings may be from such things as the operator's personal tools, gloves, etc. Other savings are in locker rooms, showers, and cafeteria facilities—all representing valuable plant space, which will not require enlarging if company growth is achieved with flexible automation systems.

Computer Integrated Manufacturing Systems (CIMS)

Computer Integrated Manufacturing (CIM) is the term used to describe the modern approach to manufacturing. Although CIM encompasses many of the other advanced manufacturing technologies such as Computer Numerical Control (CNC), Computer-Aided Design/Computer-Aided Manufacturing (CAD/CAM), robotics, and Just In Time (JIT) delivery, it is more than a new technology or a new concept. Computer Integrated Manufacturing is actually an entirely new approach to manufacturing, a new way of doing business.

To understand CIM, it is necessary to begin with a comparison of modern and traditional manufacturing. Modern manufacturing encompasses all of the activities and processes necessary to convert raw materials into finished products, deliver them to the market, and support them in the field.

Activities of CIMS

These activities include the following: ① Identifying a need for a product; ② Designing a product to meet the needs; ③ Obtaining the raw materials needed to produce the product; ④ Applying appropriate processes to transform the raw materials into finished products; ⑤ Transporting products to the market; ⑥ Maintaining the products to endure proper performance in the field.

Modern view of manufacturing can be compared with the more limited traditional view that focused almost entirely on the conversion processes. The old approach excluded such critical pre-conversion elements as market analysis research, development, and design, as well as such after-conversion elements as product delivery and product maintenance. In other words, in the old approach to manufacturing, only those processes that took place on the shop floor were considered manufacturing. This traditional approach of separating the overall concept into numerous stand-alone specialized elements was not fundamentally changed with the advent of automation.

With CIM, not only are the various elements automated, but also the islands of automation are all linked together or integrated. Integration means that a system can provide complete and instantaneous sharing of information. In modern manufacturing, integration is accomplished by computers. With this background, CIM can now be defined as the total integration of all manufacturing elements through the use of computers.

Figure 3-3 is an illustration of a CIM system, which shows how the various machines and processes used in the conversion process are integrated. However, such an illustration cannot show that research, development, design, marketing, sales, shipping, receiving, management, and production personnel all have instant access to all information generated in this system. This is what makes it a CIM system. Progress is being made toward the eventual full realization of CIM in manufacturing.

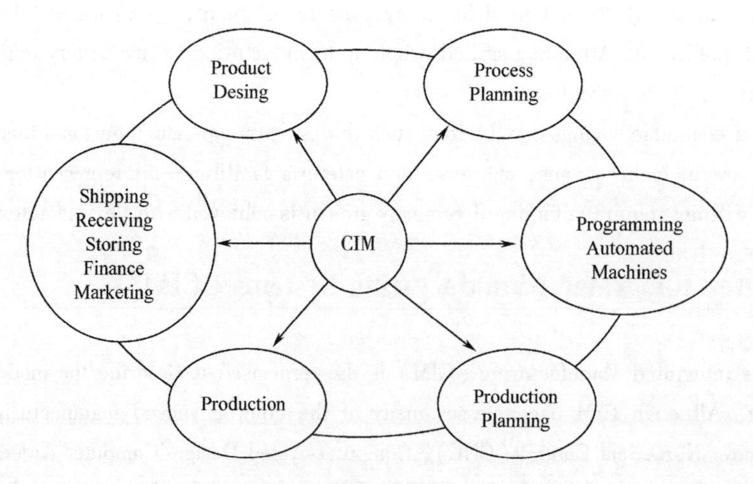

Figure 3-3 Major Components of CIM

Benefits of CIMS

Fully integrated manufacturing firms will realize a number of benefits from CIM: ① Product quali-

ty increases; ② Lead times are reduced; ③ Direct labor costs are reduced; ④ Product development times are reduced; ⑤ Inventories are reduced; ⑥ Overall productivity increases; ⑦ Design quality increased.

Lean Production (LP)

Lean production is an assembly-line manufacturing methodology developed originally for Toyota and the manufacture of automobiles. It is also known as the Toyota Production System. The goal of lean production is described as "to get the right things to the right place at the right time, the first time, while minimizing waste and being open to change". Engineer Ohno, who is credited with developing the principles of lean production, discovered that in addition to eliminating waste, his methodology led to improved product flow and better quality.

Instead of devoting resources to planning what would be required for future manufacturing, Toyota focused on reducing system response time so that the production system was capable of immediately changing and adapting to market demands.[2] In effect, their automobiles became made-to-order. The principles of lean production enabled the company to deliver on demand, minimize inventory, maximize the use of multi-skilled employees, flatten the management structure, and focus resources where they were needed.

Ten Rules of LP

The ten rules of lean production can be summarized: ① Eliminate waste; ② Minimize inventory; ③ Maximize flow; ④ Pull production from customer demand; ⑤ Meet customer requirements; ⑥ Do it right at the first time; ⑦ Empower workers; ⑧ Design for rapid changeover; ⑨ Partner with suppliers; ⑩ Create a culture of continuous improvement.

Lean production, as the name suggests involves producing goods and services while stripping out waste. As a result, there is no 'fat' involved in production.

Agile Manufacturing (AM)

Rapid, severe, and uncertain change is the most unsettling market reality that companies and people must cope with today. New products, even whole markets, appear, mutate and disappear within shorter and shorter periods of time. The pace of innovation continues to quicken, and the direction of innovation is often unpredictable. Product variety has proliferated to a bewildering degree. Agility is a comprehensive response to the challenges posed by a business environment dominated by change and uncertainty.

For a company, to be agile is to be capable of operating profitably in a competitive environment of continually and unpredictably customer opportunities.

For an individual, to be agile is to be capable of contributing to the bottom line of a company that is constantly reorganizing its human and technological resources in response to unpredictably changing customer opportunities.

But marketplace change is only one dimension of the competitive pressures that companies and people are experiencing today. At a deeper level, we are changing from a competitive environment in which mass-market products and services were standardized, long-lived, information-poor and exchanged in one-time transaction to an environment in which companies compete globally with niche market products and services that are individualized, short-lived, information-rich, and exchanged on an ongoing basis with customers.

Only those companies that respond to the deeper structural changes taking place in the commercial competition will be able to make sense of, and profit from, the superficially chaotic changes occurring at the level of the market place. A more complete definition of agility, then, is that it is comprehensive response to the business challenges of profiting from rapidly changing and continually fragmenting global markets for high quality, high performance, customer configured foods and services.

Agility is, in the end, about making money in and from a turbulent, intensely competitive business environment.

A New Manufacturing Strategy

Agility challenges the prevailing modes of organization, management, production, and competitiveness. It is explicitly strategic rather than tactical, taking no established practices for granted. Agile competition demands that the process that supports the creation, production, and distribution of goods and services be centered on the customer-perceived value of products. This is very different from building a customer-centered company enhancing the satisfaction that a customer experiences in dealing with a company adds value and can improve focus and even efficiency. But customer-centered operations are fully consistent with the mass-production mode. Centering a company on product lines that enrich customers to have for them, moves beyond the traditional mass-production systems, however efficient it may be.

Successful agile companies, therefore, know a great deal about individual customers and interact with them routinely and intensively. Neither knowledge of individual customers nor interaction on this level was relevant to mass-production-era competitors. As suppliers of standardized, uniform goods and services, mass-production-era competitors relied on market surveys that created an abstraction: the "average" or "typical" customers. However, individuality could not be accommodated in a mass-production- competitive environment.

By contrast, offering individualized products, not a bewildering list of options and models but a choice of ordering a product configured by the vendor to the particular requirements of individual customers, is the feature of agile competition, success entails formulating customer-value-added business strategies for competing in the highest-value-added markets, that is, in what are today the most profitable, and the most competitive markets.

The production operations of a successful agile company, its organizational structure, management philosophy, personnel requirements, and technology investments, are all "pulled" by these customer-opportunity-centered business strategies. In an agile competitive environment, there is no one tight way to organize and operate a company. Management can adopt the mix of multiple, concurrent strategies that will be most profitable for that company, given the variety of customers it serves and the various changing markets in which it competes. No one strategy, no one mode of organization or operation, will be successful for long. Nor will any strategy be optimal for all customers or all markets. The life expectancy of decisions that work for a company at any particular time will depend on the rate of change

in the markets in which that company competes, but it will always by far shorter than companies are used to nowadays.

AM Is a Production Mode of the 21st Century

The concept of Agile Manufacturing was introduced in 1991 by a government-sponsored research effort at Lehigh University. This new project, sponsored by the Advanced Research Project Agency and the U. S. Air Force, and in partnership with Lehigh, aims to "put meat on the bones" of the Agile concept articulated at Lehigh.

The Agile research team is using six tools in its work: transactions analysis, activity/cost chains, organization maps, key characteristics, contact chains, and Agility Metrics. Some of these are new while others are extensions of existing research techniques or adaptations of methods being used in industry already. Along with many of these tools the team is developing pictorial ways of capturing the information. Called "maps", each map shows one view of the physical, organizational, informational, or engineering information being shared by web participants. These maps are proving useful in understanding the web environment. No single map, however, seems able to show the whole situation.

The brief explanations of the tools used in this study are:

- Transactions analyses are interview-based studies of how organizations operate;
- Activity/cost chains are extension of activity-based costing;
- Organization maps show explicitly who does what in the web of suppliers;
- Key characteristics are currently in use at companies. Key characteristics are aspects of the product that require close attention. They are intended to capture customer requirements and express them systematically as design and production metrics;
- Contact chains link the key characteristics of assemblies of parts and fixtures to each other so as to describe how fit up is suppose to be achieved;
- Agility Metrics are intended to help companies determine if they are operating in an agile way.

As a result, the vision of agility having been for many years is that of a deliberate, competitive response to the constantly changing markets. As a comprehensive system, agility defines a new paradigm for doing business. It reflects a new mind-set about making, selling, and buying, openness to new forms of commercial relationships and new measures for assessing the performance of the companies and people.

E-Manufacturing

In the past decade, so much ink has been spilled (not to mention blood and treasure) on the concepts of e-business and e-manufacturing that it has been extraordinarily difficult to separate hope from hype. If the early pronouncements from e-seers were to be believed, the Internet was destined to become a force of nature that, within only a few years, would transform manufacturers and manicuring processes beyond all recognition. Everyone—customers, suppliers, management, line employees, machines, etc. —would be on-line, and fully integrated. It would be a grand alignment—one that would convert a customer's every e-whim into perfectly realized product, with all customer communication and transactions handled via the web, products designed collaboratively with customers on-line, all the right inputs delivered in exactly the right quantities at exactly the right millisecond (cued, of course, over the web), machines in production across the planet conversing with each other in a web-enabled symphony of synchronization, and total process transparency of all shop floors to the top floor, making managers omniscient gods of a brave new manufacturing universe.

These initial predictions now seem overly rosy at best, yet it is far too easy (and unfair) to dismiss e-business and e-manufacturing as fads in the same category as buying pet food and barbeque grills over the Internet. Gartner Group has estimated that as of 2001, only 1 percent of U. S. manufacturers had what could be considered full-scale e-manufacturing implementations. By 2006, U. S. Dept. of Commerce has estimated that almost half of the U. S. workforce will be employed by industries that are either major producers or intensive users of information technology products and services. The most successful e-companies, it turns out, have not been companies with ". com" emblazoned after their name, but, rather, traditional powerhouses like Intel and General Electric, who have led the way on everything from selling goods and services over the web to Internet-enabled core manufacturing processes. Perhaps most startlingly, the U. S. Bureau of Labor Statistics has projected that the rise of e-manufacturing could potentially equal or even exceed the impact of steam and electricity on industrial productivity. The Bureau recently concluded that the application of computers and early uses of the Internet in the supply chain had been responsible for a 3-percent point increase in annual U. S. manufacturing productivity growth, to 5 percent, during the 1973 – 1999 timeframe. The Bureau then projected that the rise of e-manufacturing could build upon those gains by boosting productivity growth by another two percentage points to an astounding 7 percent per year. In fact, many analysts have pointed to e-manufacturing as the next true paradigm shift in manufacturing processes—albeit one that sill take a long time to fulfill, but one that will ultimately be so pervasive that the term "e-manufacturing" will eventually become synonymous with manufacturing itself.

What Is E-Manufacturing?

As is common with new phenomena, there is currently a good deal of semantic confusion around the words "e-business" and "e-manufacturing". Let us therefore start with some simple working definitions. "E-business" is defined most cogently and accurately as the application of the Internet to business. Somewhat confusingly, "e-business" is sometimes characterized as synonymous with "e-commerce", which is more narrowly defined as the buying and selling of things on the Internet. "e-commerce" is just one subset of "e-business" —and one, though it dominated e-business-related headlines during the go-go nineties, will ultimately be one of the less important applications of the internet to business. Far more important than whether one can buy things on the Internet is the question of whether the Internet, like electricity and other fundamental technologies, can actually change (a) the fundamental customer value produced by business and (b) the efficiency via which that value can produced.

This is where "e-manufacturing" comes in. E-manufacturing can be most cogently and generally described as the application of the Internet to manufacturing. Let us first say what it is not, for the sake of analytical clarity: e-manufacturing as a discipline is not the same thing as production automation or the so-called "digital factory". The application of computing technology to the factory floor is its

own phenomenon, and can be pursued wholly independently of any use of the Internet. That being said, while e-manufacturing is not the same thing as production automation, it is perfectly complementary to the idea of production automation—an additional strategy and approach that can turbocharge the value produced by the application of technology to the factory.

What Is the Future of E-Manufacturing?

A realistic projection of the future of e-manufacturing would simultaneously take into account the very real power of the innovations embodied in the e-manufacturing paradigm, while also noting the fundamental difficulties of changing any manufacturing culture. While the good news is that approaches and technologies have finally arrived that can help make e-manufacturing a reality, and that companies across multiple industries have made enormous gains through e-manufacturing, it is nevertheless the case that an e-manufacturing implementation remains an exercise in organizational change as much as technological change—and organizational change is never easy. However, there is much to be gained from careful analysis of one's manufacturing enterprise, and applying the frameworks of e-manufacturing to see where value can be produced. It is a concept that may have as much impact on manufacturing value as the motions of mass production, total quality management (TQM), and lean manufacturing have had, and it is certainly beneficial for every manufacturing engineer to be knowledgeable about its fundamental principles and goals.

Notes

1. Future FMS will contain many manufacturing cells, each cell consisting of robot serving several computers NC (CNC) machine tools or other stand alone systems such as an inspection machine, a welder, an Electron Discharge Machining (EDM) machine, etc.

句意：未来的柔性制造系统将包括许多制造单元，每个制造单元都有机器人服务于几个计算机数字控制的加工机床，或其他独立的系统，如一个检测机、一个焊接机、一个电火花机床等。

2. Instead of devoting resources to planning what would be required for future manufacturing, Toyota focused on reducing system response time so that the production system was capable of immediately changing and adapting to market demands.

句意：丰田公司放弃了致力于对未来制造业至关重要的资源规划，取而代之的是将焦点放在缩短系统的响应时间上，以使生产系统能快速变化以适应市场的需求。

Exercises

1. What are the main functions in a computer control system?

2. What is Group Technology (GT)? And what are the major tasks to application of Group Technology?

3. What is Flexible Manufacturing System (FMS)? And how many typical parts does it have? What are the advantages of FMS?

4. What do you think of Agile Manufacturing (AM) is a Production Mode of the 21st Century?

5. In your opinion what is the future of e-manufacturing?

Unit 3 Manufacturing Support Systems

The need for illustrating, visualizing, and documenting mechanical designs prior to production has exited ever since human beings began creating machines, mechanisms, and products. Over the last century, methods for achieving this function have evolved dramatically from the blackboard illustrations of the early twentieth century and from manual drafting systems that were commonplace 50 years ago to today's automated 3D solid modeling software. As computer technology has advanced, so have the tools designers and product engineers use to create and illustrate design concepts. Today, powerful computer hardware and software have supplanted the drafting tables and T-squares of the 1960s and have advanced to the point of playing a pivotal role in not only improving design visualization but also in driving the entire manufacturing process.

CAD (Computer-Aided Design)

The advances made over the last 30 years in the use of computers for mechanical design have occurred at a more rapid pace than all the progress in design visualization that preceded the advent of computer technology. When computer hardware and software systems first appeared, the acronym CAD actually represented the term computer-aided drafting. That's because the early 2D computer design packages merely automated the manual drafting process. The first 2D CAD packages enabled designers/drafters to produce design drawings and manufacturing documentation more efficiently than the manual drafting of the past. The introduction of 2D drafting packages represented the first widespread migration of engineers to new design tools, and manufacturers readily embraced this technology because of the productivity gains it offered.

The next stage in the evolution of design tools was the move from 2D to 3D design systems. Beginning in the 1990s, this represented the second large migration of engineers to a new design paradigm and the watershed that shifted the meaning of the acronym CAD from "computer-aided drafting" to "computer-aided design". That's because 3D solid modeling removed the emphasis from using computer technology to document or capture a design concept and gave engineers a tool that truly helped them create more innovative designs and manufacture higher quality products. [1]

Instead of having the production of an engineering drawing as the final goal, engineers employ the enhanced design visualization and manipulation capabilities of 3D CAD systems to refine designs, improve products, and create 3D design data, which can be leveraged throughput the product development process.

Not all 3D solid modelers are the same, and since the introduction of 3D CAD systems two decades ago, many advances have been made. The early 3D systems were slow, expensive, based on proprietary hardware, and difficult to use because they frequently required the memorization of manual commands. The introduction of affordable, yet powerful computers and the Windows operating environment gave 3D CAD developers the foundation they needed to create the 3D CAD technology enable the

following benefits:

- Parametric Design. All features and dimensions are driven off (by) design parameters. When an engineer wants to change the design, he or she simply changes the value of the parameter, and the geometry updates accordingly;
- Bidirectional associability means quick design changes;
- Intelligent Geometry for Downstream Applications;
- Large Assembly Capabilities;
- Configurations of Derivative Products;
- Design Stylization;
- Automatic Creation of Drawings;
- Communications Design Intent;
- Assesses Fit and Tolerance Problems;
- Minimizes Reliance on Physical Prototyping;
- Eliminates Lengthy Error Checking.

Today's 3D CAD systems are the culmination of more than 30 years of research and development and have demonstrated and proven their value in actual manufacturing environments for more than a decade. 3D systems are much better than their predecessors in capturing and communicating the engineer's original design intent and automating engineering tasks, creating a sound platform on which the entire manufacturing process is based.

CAM (Computer-Aided Manufacturing)

Just as the acronym CAD evolved from meaning "computer-aided drafting" to "computer-aided design", the meaning of the acronym CAM has changed from "computer-aided machining" to "computer-aided manufacturing". The basic premise of CAM technology is to leverage product design information (CAD data) to drive manufacturing functions. The development of CAM technology to automate and manage machining, tolling, and mold creation with greater speed and accuracy if intimately linked to the development of CAD technology, which is why the term CAD/CAM is often used as a single acronym.

The introduction of CAM systems allowed manufacturing and tooling engineers to write computer programs to control machine tool operations such as milling and turning. These computer numerically controlled (CNC or NC) programs contain hundreds or thousands of simple commands, much like driving instructions, needed to move the machine tool precisely from one position to the next. These commands are sent to the machine tool's controller to control highly precise stepper motors connected to the machine tool's various axed of travel. CNC control represents a huge improvement over the traditional method of reading a blueprint and manually adjusting the position of a machine tool through hand cranks. The accuracy and repeatability of CNC machining has had a permanent impact on the reliability and quality of today's manufacturing environment.

With the development of 3D CAD solid modeling systems, the interim step of developing computer code to control 3D CAM machining operations has been automated. Because the data included in solid models represent three-dimensional shapes with complete accuracy, today's CAM systems can directly import 3D solid models and use them to generate the CNC computer code required to control manufacturing operations with an extremely high degree of precision. While manufacturers initially applied CAM technology for tooling and mass production machining operations, its use has expanded to include other manufacturing processes such as the creation of molds for plastic injection-molding and certain automatic (robotic) assembly operations, all directly from the 3D solid model.

CAPP (Computer-Aided Process Planning)

There is much interest by manufacturing firms in automating the task of process planning using CAPP systems. The shop-trained people who are familiar with the details of machining and other processes are gradually retiring, and these people will be unavailable in the future to do process planning. An alternative way of accomplishing this function is needed, and CAPP systems are providing this alternative. CAPP is usually considered to be part of computer-aided manufacturing (CAM). However, this tends to imply that CAM is a stand-alone system. In fact, a synergy result that CAM is combined with computer-aided design creates a CAD/CAM system. In such a system, CAPP becomes the direct connection between design and manufacturing. The benefits derived from computer-automated process planning include the following:

- Process rationalization and standardization. Automated process planning leads to more logical and consistent process plans than when process planning is done completely manually. Standard plans tend to result in lower manufacturing costs and higher product quality;
- Increased productivity of process planners. The systematic approach and the availability of standard process plans in the data files permit more work to be accomplished by the process planners;
- Reduced lead time for process planning. Process planners working with a CAPP system can provide route sheets in a shorter lead time compared to manual preparation;
- Improved legibility. Computer-prepared route sheets are neater and easier to read than manually prepared route sheets;
- Incorporation of other application programs. The CAPP program can be interfaced with other application programs, such as cost estimating and work standards.

CAE (Computer-Aided Engineering)

CAE stands for computer-aided engineering and primarily encompasses two engineering software technologies that manufacturers use in conjunction with CAD to engineer, analyze, and optimize product designs. Design analysis and knowledge-based engineering (KBE) applications help manufacturers to refine design concepts by simulating a product's physical behavior in its operating environment and infusing the designer's knowledge and expertise into the manufacturing process. [2]

Creating a CAD model is one thing, but capturing the knowledge or design intent that went into designing the model and reusing it as part of manufacturing represents the essence of KBE software

technology. These applications propagate the designer's process-specific knowledge throughout the designing and manufacturing process, leveraging the organization's knowledge to produce consistent quality and production efficiencies.

Design analysis, also frequently referred to as finite element analysis (FEA), is a software technology used to simulate the physical behavior of a design under specific conditions. FEA breaks a solid model down into many small and simple geometric elements (bricks, tetrahedrons) to solve a series of equations formulated around how these elements interact with each other and the external nodes. Using this technique, engineers can simulate responses of designs to operating forces and use these results to improve design performance and minimize the need to build physical prototypes.

Some of the questions FEA help answer early in the design include:

- Will structural stresses or fatigue cause it to break, buckle, or deform?
- Will thermal stresses weaken a component or cause it to fail?
- Will electromagnetic forces cause a system to behave in a manner inconsistent with its intended use?
- How much material can be removed, and where, while still maintaining the required safety factor?

Depending on the component of system, and how it is used, effective design analysis can mean the difference between product success and acceptance, or even life and death. For example, airplane manufacturers use FEA to ensure that aircraft wings can withstand the forces of flight. The more common types of design analyses that are performed include structural, thermal, kinematics (motion), electromagnetic, and fluid dynamics analyses.

Early design analysis software packages were separate applications, often with their own geometric modeling application. Analysts would often have to rebuild a model after they received it from a designer in order to complete the particular type of design analysis required. Today, many analysis systems operate directly on the 3D CAD solid model, and some packages are even integrated directly with 3D CAD software, combining design and analysis within the same application. The benefit of an integrated design analysis package is that a design engineer can optimize the design as part of the conceptual design rather than after the design concept is finished.

Once a separate application, CAD has steadily evolved to work with a variety of other software applications (Figure 3-4). Product development is not a single step but rather a continuous process that impacts various departments and functions from the idea stage all the way through the actual introduction of a product. Instead of treating design as a separate, autonomous function, CAD vendors recognize that the value of 3D CAD data extends far beyond conceptual design. By making CAD data compatible with other functions such as manufacturing, documentation, and marketing, CAD developers have accelerated the rate at which information is processed throughout the product development organization and have produced deficiencies and productivity improvements that were unanticipated and unsupported during the early application of CAD technology. Increasingly, CAD data has become the data thread that weaves its way across the extended manufacturing enterprise, accelerating the rate at which information is processed throughout product development.

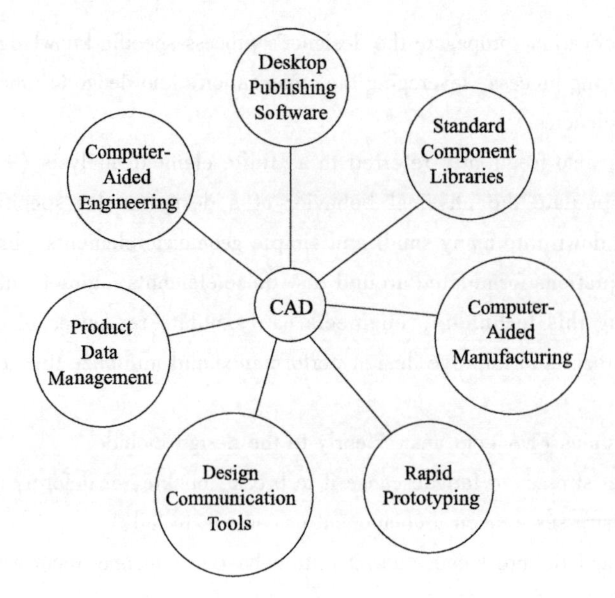

Figure 3-4 CAD Data Drives and Interacts with a Broad Array of Tools and Technologies

Notes

1. That's because 3D solid modeling removed the emphasis from using computer technology to document or capture a design concept and gave engineers a tool that truly helped them create more innovative designs and manufacture higher quality products.

句意：这是因为三维实体建模把重点从使用计算机技术转移到创建文档或捕捉设计概念上，给工程师们一个真正帮助他们开发更具创造性的设计和制造更高质量的产品的工具。

2. Design analysis and knowledge-based engineering (KBE) applications help manufacturers to refine design concepts by simulating a product's physical behavior in its operating environment and infusing the designer's knowledge and expertise into the manufacturing process.

句意：设计分析和基于知识的工程应用帮助制造者借助于模拟产品在使用环境下的物理行为来完善设计概念，并将设计者的知识和专长融入制造过程中。

Exercises

1. How do you think about the tendency of Manufacturing Systems in the coming future?

2. How many Manufacturing Support Systems do you know? And how many do you often used in your work?

3. In your opinion, which Manufacturing Support Systems have a bright future?

4. What is the CAD/CAM system?

5. How the CAPP can make the benefit for a Manufacturing System?

Chapter 4
Service Systems

Unit 1 Introduction to Service Systems

Definitions of Service

Many definitions of service are available but all contain a common theme of intangibility and simultaneous consumption. The following represents a sample definition of service:

Services are deeds, processes, and performances.

A service is an activity or series of activities of more or less intangible nature that normally, but not necessarily, take place in interactions between customer and service employees and/or physical resources or goods and/or systems of the service provider, which are provided as solutions to customer problems.

Most authorities consider that the services sector including all economic activities whose output is not a physical product or construction, is generally consumed at the time it is produced, and provides added value in forms (such as convenience, amusement, timeliness, comfort or health) that are essentially intangible concerns of its first purchaser.

A precise definition of goods and services should distinguish them on the basis of their attributes. Goods are tangible physical objects or products that can be created and transferred; they are existent over time and thus can be created and used later. A service is intangible and perishable. It is an occurrence or process that is created and used simultaneously or nearly simultaneously. While the consumer cannot retain the actual service after it is produced, the effect of the service can be retained.

A service is a time-perishable, intangible experience performed for a customer acting in the role of co-producer.

Nature of the Service

For many people, service is synonymous with servitude and brings to mind workers flipping hamburgers and waiting on tables. However, the service sector that has grown significantly over the past 30 years cannot be accurately described as composed only of low-wage or low-skill jobs in department stores and fast-food restaurants. Instead the fastest-growing jobs within the service sector are in finance, insurance, real estate, miscellaneous services (e. g., health, education, professional services), and retail trade. Note that job areas whose growth rates were less than the rate of increase in total jobs (i. e., less than 31. 8 percent) lost market share, even though they showed gains in their absolute numbers. The exceptions are in mining and manufacturing, which lost in absolute numbers and thus showed negative growth rates. This trend should accelerate with the end of the Cold War and the subsequent downsizing of the military and defense industry.

Today, service industries are the source of economic leadership. During the past 30 years, more than 44 million new jobs have been created in the service sector to absorb the influx of women into the workforce and to provide an alternative to the lack of job opportunities in manufacturing. The service industries now account for approximately 70 percent of the national income in the United States. Given that there is a limit to how many cars a consumer cart use and how much one can eat and drink, this should not be surprising. The appetite for services, however, especially innovative ones, is insatiable. Among the services presently in demand are those that reflect an aging population, such as geriatric health care, and others that reflect a two-income family, such as day care.

The growth of the service sector has produced a less cyclic national economy. During the past four recessions in the United States, employment by service industries has actually increased, while jobs in manufacturing have been lost. This suggests that consumers are willing to postpone the purchase of products but will not sacrifice essential services like education, telephone, banking, health care, and public services such as fire and police protection.

Several reasons can explain the recession-resistant nature of services. First, by their nature, services cannot be inventoried, as is the case for products. Because consumption and production occur simultaneously for services, the demand for them is more stable than that for manufactured goods. When the economy falters, many services continue to survive. Hospitals keep busy as usual, and, while commissions may drop in real estate, insurance, and security businesses, employees need not be laid off.

Second, during a recession both consumers and business firms defer capital expenditures and instead fix up and make do with existing equipment. Thus, service jobs in maintenance and repair are created.

Roles of Service

Successful growth of the service sector will depend on innovation and skilled management that will promote an ethic of continuous improvement in both quality and productivity.

The product development model that is driven by technology and engineering could be called a push theory of innovation. A concept for a new product germinates in the laboratory with a new scientific discovery that becomes a solution looking for a problem. The 3M experience with Post-it notes is one example of this innovation process. The laboratory discovery was a poor adhesive, which found a creative use as a glue for notes to be attached temporarily to objects without leaving a mark when removed.

Information technology provides many examples of the push theory of service innovation. The growth of the World Wide Web as a place of commerce is changing the delivery of services. People can browse the Internet for every imaginable product or service from around the world. In fact, to stay competitive, many businesses may soon be required to offer new cost-effective and convenient services for customers who have home computers equipped with modems.

For a manufacturing firm, product innovation is often driven by engineering—based research, but in service firms, software engineers and programmers are the technocrats who develop new innovations. Customers interact directly in the service process; therefore, the focus on meeting customer needs drives service innovation and explains why marketing plays such a central role in service management.[1]

Service innovation also can arise from exploiting information available from other activities. For example, records of sales by auto parts stores can be used to identify frequent failure areas in particular models of cars. This information has value both for the manufacturer, who can accomplish engineering changes, and for the retailer, who can diagnose customer problems. In addition, the creative use of information can be a source of new services, or it can add value to existing services. For example, an annual summary statement of transactions furnished by one's financial institution has added value at income tax time.

We have discovered that the modem industrial economies are dominated by employment in the service sector. Just as farming jobs migrated to manufacturing in the 19th century under the driving force of laborsaving technology, manufacturing jobs in due time migrated to services. Now as we begin the new millennium, an experience economy is emerging to satisfy rising expectations for services.

Service Classifications

Concepts of service management should be applicable to all service organizations. For example hospital administrators could learn something about their own business from the restaurant and hotel trade. Professional services such as consulting, law, and medicine have special problems because the professional is trained to provide a specific clinical service (to use a medical example) but is not knowledgeable in business management. Thus, professional service firms offer attractive career opportunities for many college graduates.

A service classification scheme can help to organize our discussion of service management and break down the industry barriers to shared learning. As suggested, hospitals can learn about housekeeping from hotels. Less obviously, dry-leaning establishments can learn from banks—cleaners can adapt the convenience of night deposits enjoyed by banking customers by providing laundry bags and after-hours drop-off boxes. For professional firms, scheduling a consulting engagement is similar to planning a legal defense or preparing a medical team for open-heart surgery.

To demonstrate that management problems are common across service industries, Roger Schmenner proposed the service process matrix in Figure 4-1. In this matrix, services are classified across two dimensions that significantly affect the character of the service delivery process. The vertical dimension measures the degree of labor intensity, which is defined as the ratio of labor cost to capital cost. Thus capital-intensive services such as airlines and hospitals are found in the upper row because of their considerable investment in plant and equipment relative to labor costs. Labor-intensive services such as schools and legal assistance are found in the bottom row because their labor costs are high relative to their capital requirements.

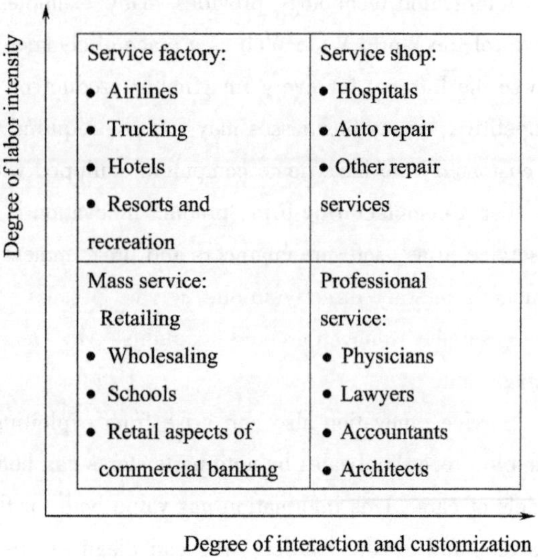

The horizontal dimension measures the degree of customer interaction and customization, which is a marketing variable that describes the ability of the customer to affect personally the nature of the service being delivered.[2] Little interaction between customer and service provider is needed when the service is standardized rather than customized. For example, a meal at McDonald's, which is assembled from prepared items, is low in customization and served with little interaction occurring between the customer and the service providers. In contrast, a doctor and patient must interact fully in the diagnostic and treatment phases to achieve satisfactory results. Patients also expect to be treated as individuals and wish to receive medical care that is customized to their particular needs. It is important to note, however, that the interaction resulting from high customization creates potential problems for management of the service delivery process.

Degree of labor intensity (vertical axis)

Service factory:	Service shop:
• Airlines • Trucking • Hotels • Resorts and recreation	• Hospitals • Auto repair • Other repair services
Mass service: Retailing • Wholesaling • Schools • Retail aspects of commercial banking	Professional service: • Physicians • Lawyers • Accountants • Architects

Degree of interaction and customization (horizontal axis)

Figure 4-1 The Service Process Matrix

The four quadrants of the service process matrix have been given names, as defined by the two dimensions, to describe the nature of the services illustrated. Service factories provide a standardized service with high capital investment, much like a line-flow manufacturing plant. Service shops permit more service customization, but they do so in a high-capital environment. Customers of a mass service will receive an undifferentiated service in a labor-intensive environment, but those seeking a professional service will be given individual attention by highly trained specialists.

Managers of services in any category, whether service factory, service shop, mass service, or professional service, share similar challenges, as noted in Figure 4-2. services with high capital requirements (i. e., low labor intensity), such as airlines and hospitals, require close monitoring of technological advances to remain competitive. This high capital investment also requires managers to schedule demand to maintain utilization of the equipment. Alternatively, managers of highly labor-intensive services, such as medical or legal professionals, must concentrate on personnel matters.

The degree of customization affects the ability to control the quality of the service being delivered and the perception of the service by the customer. Approaches to addressing each of these challenges are topics that will be discussed later.

Figure 4-2 Challenges for Service Managers

Characteristics of Service Operations

In services, a distinction must be made between inputs and resources. For services, inputs are the customers themselves, and resources are the facilitating goods, employee labor, and capital at the command of the service manager. Thus, to function, the service system must interact with the customers as participants in the service process. Because customers typically arrive at their own discretion and with unique demands on the service system, matching service capacity with demand is a challenge.

For some services, such as banking, however, the focus of activity is on process information instead of people. In these situations, information technology, such as electronic funds transfer, can be substituted for physically depositing a payroll check; thus, the presence of the customer at the bank is unnecessary. Such exceptions will be noted as we discuss the distinctive characteristics of service operations. It should be noted here that many of the unique characteristics of services, such as customer participation and perishability, are interrelated.

Customer Participation

The presence of the customer as a participant in the service process requires an attention to facility design that is not found in traditional manufacturing operations. That automobiles are made in a hot, dirty, noisy factory is of no concern to the eventual buyer because they first see the product in the pleasant surroundings of a dealer's showroom. The presence of the customer on-site requires attention to the physical surroundings of the service facility that is not necessary for the factory. For the customer, service is an experience occurring in the environment of the service facility, and the quality of service is enhanced of the service facility is designed from the customer's perspective. Attention to interior decorating, furnishings, layout, noise, and even color can influence the customer's perception of the service.

Compare the feelings invoked by picturing yourself in a stereotypical bus station with those produced by imagining yourself in an airline terminal. Of course, passengers are not allowed in the terminal's back office, which is operated in a factory-like environment. However, some innovative services have opened the back office to public scrutiny to promote confidence in the service (e. g., some restaurants provide a view into the kitchen, some auto repair bays can be observed through windows in the waiting area).

Taking the customer out of the process, however, is becoming a common practice. Consider retail banking, in which customers are encouraged to use telephone or computer transactions, direct deposit, and automatic-debit bill paying instead of actually traveling to the bank/moreover, the advent of Internet commerce gives new meaning to the phrase "window shopping".

Simultaneity

The fact that services are created and consumed simultaneously and, thus, cannot be stored is a critical feature in the management of services. This inability to inventory services precludes using the traditional manufacturing strategy of relying on inventory as a buffer to absorb fluctuations in demand. An inventory of finished goods serves as a convenient system boundary for a manufacturer, separating the internal operations of planning and control from the external environment. Thus, the manufacturing facility can be operated at a constant level of output that is most efficient, the factory is operated as a closed system, with inventory decoupling the productive system from customer demand, services, however, operate as open systems, with the full impact of demand variations being transmitted to the system.

Inventory also can be used to decouple the stages in a manufacturing process. For services, the decoupling is achieved through customer waiting. Inventory control is a major issue in manufacturing operations, whereas in services, the corresponding problem is customer waiting, or "queuing". The problems of selecting service capacity, facility utilization, and use of idle time all are balanced against customer waiting time.

The simultaneous production and consumption in services also eliminates many opportunities for quality-control intervention. A product can be inspected before delivery, but services must rely on other measures to ensure the quality of services delivered.

Perishability

A service is a perishable commodity. Consider an empty airline seat, an unoccupied hospital or hotel room, or an hour without a patient in the day of a dentist. In each case, a lost opportunity has occurred. Because a service cannot be stored, it is lost forever when not used. The full utilization of service capacity becomes a management challenge, because customer demands exhibit considerable

variation and building inventory to absorb these fluctuations is not an option.

Consumer demands for services typically exhibit very cyclic behavior over short periods of rime, with considerable variation between the peaks and valleys. The custom of eating lunch between noon and 1 p. m places a burden on restaurants to accommodate the noon rush. The practice of day-end mailing by businesses contributes to the fact that 60 percent of all letters are received at the post office between 4 and 8 p. m. The demand for emergency medical service in Los Angeles was found to vary from a low of 0. 5 calls per hour at 6 p. m. to a peak of 3. 5 calls per hour at 6 p. m. This peak-to-valley ratio of 7 to 1 also was true for fire alarms during an average day in New York City.

For recreational and transportation services, seasonal variation in demand creates surges in activity. As many students know, flights home are often booked months in advance of spring break and the Christmas holiday.

Intangibility

Services are ideas and concepts; products are things, therefore, it follows that service innovations are not patentable. To secure the benefits of a novel service concept, the firm must expand extremely rapidly and preempt any competitors. Franchising has been the vehicle to secure market areas and establish a brand name. Franchising allows the parent form to sell its idea to a local entrepreneur, thus preserving capital while retaining control and reducing risk.

The intangible nature of services also presents a problem for customers. Franchising has been the vehicle to secure market areas and establish a brand mane. Franchising allows the parent firm to sell its idea to a local entrepreneur, thus preserving capital while retaining control and reducing risk.

The intangible nature of services also presents a problem for customers. When buying a product, the customer is able to see it, feel it, and test its performance before purchase. For a service, however, the customer must rely in the reputation of the service firm. In many service areas, the government has intervened to guarantee acceptable service performances. Through the use of registration, licensing, and regulation, the government can assure consumers that the training and test performance of some service providers meet certain standards. Thus, we find that public construction plans must be approved by a registered professional engineer, a doctor must be licensed to practice medicine, and the telephone company is a regulated utility. In its efforts to "protect" the consumer, however, the government may be stifling innovation, raising barriers to entry, and generally reducing competition.

Heterogeneity

The combination of the intangible nature of services and the customer as a participant in the service delivery system results in variation of service from customer to customer. The interaction between customer and employee in services, however, creates the possibility of a more complete human work experience. In services, work activity generally is oriented toward people rather than toward things. There are exceptions, however, for services that process information (e. g., communications) of customers' property (e. g., brokerage services). In the limited customer-contact service industries, we now see a dramatic reduction in the level of labor intensiveness through the introduction of information technology.

Even the introduction of automation may strengthen personalization by eliminating the relatively routine impersonal tasks, thereby permitting increased personal attention to the remaining work. At the same time, personal attention creates opportunities for variability in the service that is provided. This is not inherently bad, however, unless customers perceive a significant variation in quality. A customer expects to be treated fairly and to be given the same service that others receive. The development of standards and of employee training in proper procedures is the key to ensuring consistency in the service provided. It is rather impractical to monitor the output of each employee, except via customer complaints.

The direct customer-employee contact has implications for service (industrial) relations as well. Autoworkers with grievances against the firm have been known to sabotage the product on the assembly line. Presumably, the final inspection will ensure that any such cars are corrected before delivery. A disgruntled service employee, however, can do irreparable harm to the organization because the employee is the firm's sole contact with customers. Therefore, the service manager must be concerned about the employees' attitudes as well as their performance. J. Willard Marriott, founder of the Marriott Hotel chain, has said, "In the service business you can't make happy guests with unhappy employees." Through training and genuine concern for employee welfare, the organizational goals can be internalized.

Notes

1. Customers interact directly in the service process; therefore, the focus on meeting customer needs drives service innovation and explains why marketing plays such a central role in service management.

句意：消费者直接参与服务过程，因此，其焦点是满足消费者的需求迫使服务发生变革，并可解释为什么行销在服务管理中占据中心地位的原因。

2. The horizontal dimension measures the degree of customer interaction and customization, which is a marketing variable that describes the ability of the customer to affect personally the nature of the service being delivered.

句意：水平坐标表示消费者参与和定制的程度，是一个描述消费者本人对服务状态影响能力的一个营销变量。

Exercises

1. What is the value of service in an economy?
2. Speculate on the effect that the Internet will have on the delivery of services.
3. Is it possible for an economy to be based entirely on services?
4. Comment on the role that marketing plays in the service innovation process.
5. What are the characteristics of services that will be most appropriate for Internet delivery?

Unit 2 Service Quality

For services, the assessment of quality is made during the service delivery process. Each customer contact is referred to as a moment of truth, an opportunity to satisfy or dissatisfy the customer. Customer satisfaction with expectations of service can be defined by comparing perceptions of service received

with expectations of service desired. When expectations are exceeded, service is perceived to be of exceptional quality—and also to be a pleasant surprise. When expectations are not met, however, service quality is deemed unacceptable. When expectations are confirmed by perceived service, quality is satisfactory. As shown in Figure 4-3, these expectations are based on several sources, including word of mouth, personal needs, and past experience.

Dimensions of Service Quality

The dimensions of service quality as shown in Figure 4-3 were identified by marketing researchers studying several different service categories: appliance repair, retail banking, long-distance telephone service, securities brokerage, and credit card companies. They identified five principal dimensions that customers use to judge service quality-reliability, responsiveness, assurance, empathy, and tangibles, which are listed in order to decline relative importance to customers.

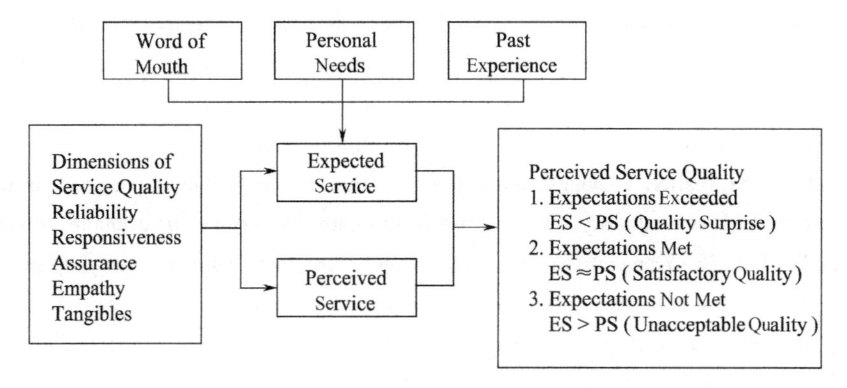

Figure 4-3 Perceived Service Quality

Reliability. The ability to perform the promised services both dependably and accurately. Reliable service performance is a customer expectation and means that the service is accomplished on time, in the same manner, and without errors every time. For example, receiving mail at approximately the same time each day is important to most people. Reliability also extends into the back office, where accuracy in billing and record keeping is expected.

Responsiveness. The willingness to help customers and to provide prompt service. Keeping customers waiting, particularly for no apparent reason, creates unnecessary negative perceptions of quality. If a service failure occurs, the ability to recover quickly and with professionalism can create very positive perceptions of quality. For example, serving complimentary drinks on a delayed flight can turn a potentially poor customer experience into one that is remembered favorably.

Assurance. The knowledge and courtesy of employees as well as their ability to convey trust and confidence. The assurance dimension includes the following features: competence to perform the service, politeness and respect for the customer, effective communication with the customer, and the general attitude that the server has the customer's best interests at heart.

Empathy. The provision of caring, individualized attention to customers. Empathy includes the following features: approachability, sensitivity, and effort to understand the customer's needs. One example of empathy is the ability of an airline gate attendant to make a customer's missed connection the attendant's own problem and to find a solution.

Tangibles. The appearance of physical facilities, equipment, personnel, and communication materials. The condition of the physical surroundings (e. g., cleanliness) is tangible evidence of the care and attention to detail that is exhibited by the service provider. This assessment dimension also can extend to the conduct of other customers in the service (e. g., a nosy guest in the nest room at a hotel).

Customers use these five dimensions to form their judgments of service quality, which are based on a comparison between expected and perceived service. The gap between expected and perceived service is a measure of service quality; satisfaction is either negative or positive.

Measuring of Service Quality

Measuring service quality is a challenge because customer satisfaction is determined by many intangible factors. Unlike a product with physical features that call be objectively measured (e. g., the fit and finish of a can service quality contains many psychological features (e. g., the ambiance of a restaurant). In addition, service quality often extends beyond the immediate encounter because, as in the case of health care. It has an impact on a person's future quality of life.

Scope of service quality. A comprehensive view of the service system is necessary to identify the possible measures of service quality. We will use health care delivery as our example service, and we will view quality from five perspectives: content, process, structure, outcome, and impact. For health care, the scope of service quality obviously extends beyond the quality of care that is comprehensive view of service quality need not be limited to health care, however, as demonstrated by the negative economic impact of failed savings-and-loan institutions on their customers as well as on the community as a whole.[1]

Contents. Are standard procedures being followed? For example, is the dentist following accepted dental practices when extracting a tooth? For routine services, standard operating procedures generally are developed, and service personnel are expected to follow these established procedures. In health care, a formal peer-review system, called Professional Standards Review Organization (PSRO), has been developed as a method of self-regulation. Under this system, physicians in a community or specialty establish standards for their practices and meet regularly to review peer performance so that compliance is assured.

Process. Is the sequence of events in the service process appropriate? The primary concern here is maintaining a logical sequence of activities and a well-coordinated use of service resources. Interactions between the customer and the service personnel are monitored. Also of interest are the interactions and communications among the service workers. Check sheets are common measurement devices. For emergency services such fire and ambulance, disaster drills in a realistic setting are used to test a unit's performance; problems with coordination and activity sequencing can be identified and corrected through these practice sessions.

Structure. Are the physical facilities and organizational design adequate for the service? The physical facilities and support equipment are only part of the structural dimension, however. Qualifications

of the personnel and the organizational design also are important quality dimensions. For example, the quality of medical care in a group practice can be enhanced by an on-site laboratory and x-ray facilities. More important, the organization may facilitate consultations among the participating physicians. A group medical practice also provides the opportunity for peer pressure to control the quality of care that its members provide.

Adequacy of the physical facilities and equipment can be determined by comparison with set standards for quality conformance. One well-known fast-food restaurant is recognized for its attention to cleanliness. Store managers are subjected to surprise inspections in which they are held responsible for the appearance of the parking lot, sidewalk, and restaurant interior. Personnel qualifications for hiring, promotion, and merit increases also are matters of meeting standards. University professors seldom are granted tenure unless they have published, because the ability to publish in a refereed journal is considered to be independent evidence of research quality. [2] A measure of organizational effectiveness in controlling quality would be the presence of active self-evaluation procedures and members' knowledge of their peers' performances.

Outcome. What change in status has the service effected? The ultimate measure of service quality is a study of the end result. Is the consumer satisfied? We are all familiar with the cards on restaurant tables that request our comments on the quality of service. Complaints by consumers are one of the most effective measures of the quality of service. Complaints by consumers are one of the most effective measures of the quality outcome dimension. For public services, the assumption often is made that the status quality is acceptable unless the level of complaints begins to rise. The concept of monitoring output quality by tracking some measure (e. g., the number of complaints) is widely used. For example, the performance of a hospital is monitored by comparing certain measures against industry norms. The infection rate per 1,000 surgeries might be used to identify hospitals that may be using substandard operating room procedures.

Clever approaches to measuring outcome quality often are employed. For example, the quality of trash pickup in a city can be documented by taking pictures of the city streets after the trash vehicles have made their rounds. One often-forgotten measure of outcome quality is the satisfaction of empowered service personnel with their own performance.

Impact. What is the long-range effect of the service on the consumer? Are the citizens of a community able to walk the streets at night with a sense of security? The result of a poll asking that question would be a measure of the impact of police performance. The overall impact of health care often is measured by life expectancy or the infant mortality rate, and the impact of education often is measured by literacy rates and performance on nationally standardized tests.

It should be noted, however, that the impact also must include a measure of service and accessibility, which usually is quoted as the population served per unit area. Health care in the United States is criticized for the financial barriers to patient accessibility in general but especially in rural and large inner-city areas. As a result, this country's impact measures of life expectancy and infant mortality are far worse than those in all other industrial countries and even several Third World countries. In a similar fashion, the literacy rate is a measure of the impact of the education system, and again, the United States lags behind many other nations. Health care and education are perhaps the two most essential services in the United States today. Clearly, they are in great need of managers who can devise and implement excellent and innovative service operations strategies.

A commercial example of an impact measurement is the number of hamburgers sold, which once was displayed in neon lights by McDonald's. In addition, a bank's lending rate for minorities could be a measure of that institution's economic impact on a community.

Notes

1. For health care, the scope of service quality obviously extends beyond the quality of care that is comprehensive view of service quality need not be limited to health care, however, as demonstrated by the negative economic impact of failed savings-and-loan institutions on their customers as well as on the community as a whole.

句意：然而，就医疗卫生行业而言，服务质量的范围明显地超出了护理的质量范围。从更大范围来看，服务质量不再局限于医疗卫生，正如由失败的存-贷机构对他们的客户以及整个群体所带来的负面经济影响所表明的一样。

2. University professors seldom are granted tenure unless they have published, because the ability to publish in a refereed journal is considered to be independent evidence of research quality.

句意：大学教授如果不发表论著很少能被聘用，因为在期刊上发表论文的能力被认为是独立研究水平的证据。

Exercises

1. What is the management problem associated with allowing service employees to exercise judgment in meeting customer needs?

2. What factors are important for a manager to consider when attempting to enhance a service form's image?

3. State the "distinctive characteristics of service operations" for a service with which you are familiar.

4. In your opinion, what is the good service for you and how to measure the quality of service?

5. Do you have the experience that you complain the service quality for your product and what is the result at last?

Chapter 5
Production Planning and Control

Unit 1 The Main Idea of Production Planning

Production and Production Process

Production planning and control belongs to production and operations management category, and it is the core of the production and management. Production and operations management is based on the management of production system or production process. Therefore, in order to study production planning and control, we must understand what are production processes and the construction of the processes.

Production

Production is the creation of goods and services. Activities creating goods and services take place in all organizations. In manufacturing firms, the production activities that create goods are usually quite obvious. In them, there are tangible products can be creation such as an automobile or a computer. Among organizations that do not create physical products, often when services are performed, intangible goods are produced, such as the education of students. Regardless of whether the end product is goods or services, the production activities that go on in the organization are often needs planning and control.

Production Processes

A process is any activity or a group of activities that takes one or more inputs, transforms and adds value to them, and provides one or more outputs for its customers. The types of processes may vary. For example, at a factory a primary process would be a physical or chemical change of raw materials into products. But there also are many nonmanufacturing processes at a factory, such as order fulfillment, making due-date promises to customers, and inventory control.[1] At an airline, a primary process would be the movement of passengers and their luggage from one location to another, but there are also processes for making reservations, checking in passengers, and scheduling crews.

As illustrated in Figure 5-1, processes have inputs and customer outputs. Inputs include human resources (workers and managers), capital (equipment and facilities), purchased materials and services, land, and energy. The numbered circles represent operations through which services, products, or customers pass and where processes are performed. The arrows represent flows and can cross because one job or customer can have different requirements (and thus, a different flow pattern) than the next job or customer. Processes provide outputs—often services (which can take the form of information) —to their "customers." Both manufacturing and service organizations now realize that every process and every person in an organization has customers. Some are external customers, who may be either end users or intermediaries (such as manufacturers, wholesalers, or retailers) buying the firm's finished products and services. Others are internal customers who may be one or more other employees who rely on inputs from earlier processes in order to perform processes in the next office, shop, or department. Either way, processes must be managed with the customer in mind.

Figure 5-1 can represent a whole firm, a department or small group, or even a single in dividable. Each one has inputs and uses processes at various operations to provide outputs. The dashed lines represent two special types of input: participation by customers and in formation on performance from both internal and external sources. Participation by customers occurs not only when they receive outputs but also when they take an active part in the processes, such as when students participate in a class discussion. Information on performance includes internal reports on customer service or inventory levels and external information from market research, government reports, or telephone calls from suppliers. Managers need all types of information to manage processes most effectively.

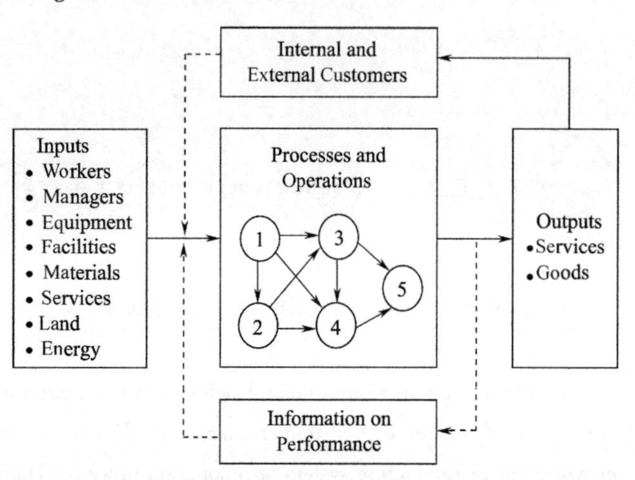

Figure 5-1 Processes and Operations

Introduction to Production Planning

Production planning refers to the preparation of specific production plan work. It will determine the production plan through a series of comprehensive balance works. The reason for designing the production planning system is to continuously improve the level of production planning, and provide an optimized production plan for the operation of industrial production systems. The so-called optimized production plan is beneficial to take advantage of sales opportunities to meet market demand; beneficial to take advantage of profit opportunities and achieve the lowest cost of production; to make full use of production resources, minimize waste and limitation of the production resources. [2]

The steps of production planning

(1) Preparation for production planning.

First of all, predicting market demand in planning period, accounting company's own production

capacity and providing an external needs and internal possibility basis

(2) To determine the targets of the plan.

Secondly, according to the principle that is making full use of various resources and increasing economic benefits to meet the social needs, on the basis of overall balance to determine targets of the production planning such as product variety, quality, yield and value indicators which should be finished in the planning period.

(3) Account and balance the production capacity.

Third, according to account production capacity and balance implementation of the planning task, while make full use of the production capacity to achieve the best economic benefits.

(4) Determine product production schedule.

Next is to arrange product production schedule properly, in which ensures the realization of production targets in time, and also ensures the production have a nice tie with marketing and balanced production, so that the production capacity is fully utilized and to reduce production costs.

(5) Organized and checked of the implementation of production planning.

After the completion of production planning, the key to achieve prospective objectives is to organize implementation of the plan. During the implementation, checking the implementation in time, tracking the production process, analyzing and evaluating, summing up experience and to correct deviations are necessary. Based on above all works, the last thing is to table and fill in those production schedules.

Production Planning System of Enterprise

Production planning can be divided into long-term plan, medium-term plan and short-term plan. And between these three plans are closely linked and coordinated to constitute the total system of production planning system.

Long-Term Plan

The length of a long-term plan is generally 3-5 years. It is the planning about the major problems of the enterprise in the production, technology, financial and other aspects. It includes product and market development plans, resource development plans and production strategies and financial plans and other types of plans. It specifies the long-term plan, the first thing is to make analysis with the economic, technological, political environment, and then to predict of the business development, and to determine the overall development goal of enterprises, such as production, output value, profits, quality, variety and other aspects of growth and the level should reached.

Medium-Term Plan

The length of a medium-term plan is generally one year or longer period of time. It is the usual annual production plan. The medium-term plan includes two plans: production planning outline (PP) and the schedule of products production (or Master production schedule, MPS).

Production Planning Outline

Production planning outline (PP) provides the production targets of the enterprise within the program year. Those targets are represented by a series of indicators to require the standards that the businesses should reached in variety, quality, production and value, etc.

Master Production Scheduling

Master production schedule (MPS) is the to take production plan outline into concrete annual production of monthly yield plans according to product varieties and specifications. MPS typically compiled once every half a year, also can be pressed scrolling update for a shorter period of time. After specifying the product production schedule, you still need to account for the balance of production capacity to ensure the feasibility of the plan. But at this level, production capacity accounting and balancing are rough, so it is rough cut capacity planning (RCCP). Of course, at the same time of the checking with production capacities, we should also check the supply capacities of other resources, such as the supply and demand balance situation of raw materials, energy, purchased parts, transportation and others. We will describe the MPS in Unit 2.

Short-Term Plans

The length of a short-term plan is less than six months, which usually are months or cross monthly plan, include material requirements planning (MRP), capacity requirements plan (CRP), and final assembly plan and operation schedule and control work in the workshop during the implementation process of these plans.

The composition of production plan and relationship between various plans are illustrated in the Figure 5-2.

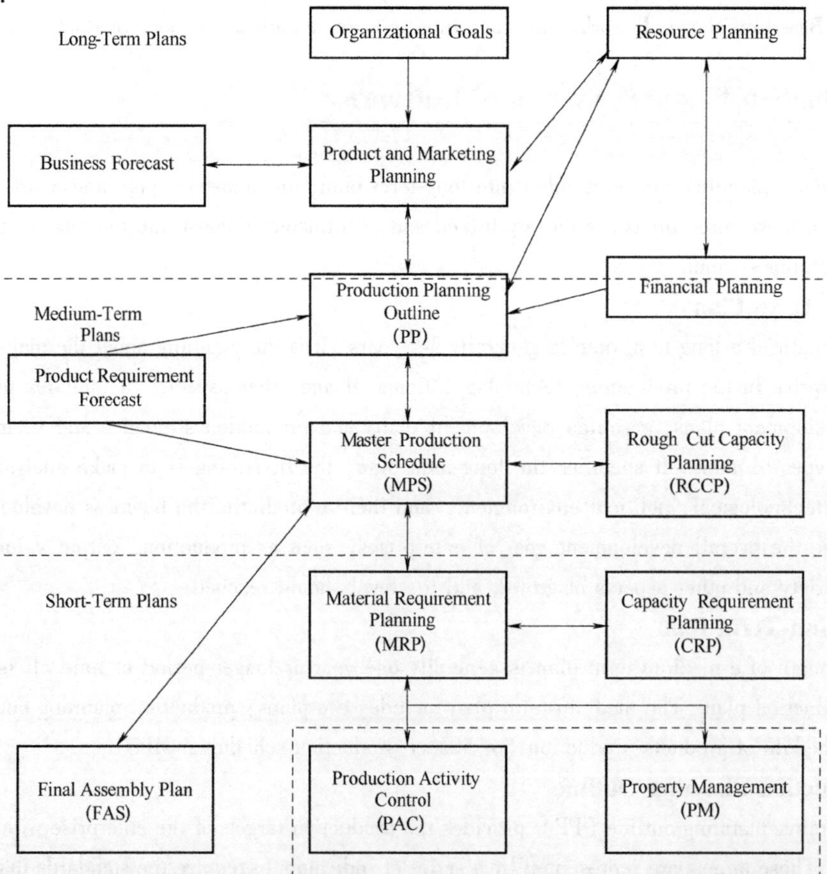

Figure 5-2 The Composition of Production Plan and Relationship between Various Plans

Notes

1. For example, at a factory a primary process would be a physical or chemical change of raw materials into products. But there also are many non-manufacturing processes at a factory, such as order fulfillment, making due-date promises to customers, and inventory control.

句意：例如，在一个工厂，主要的生产过程可能是将原材料经过物理或化学变化变成产品的过程。但是，在生产过程中还包括许多非生产过程，如完成订单，顾客到期票据的及时处理，以及库存控制。

2. The so-called optimized production plan is beneficial to take advantage of sales opportunities to meet market demand; beneficial to take advantage of profit opportunities and achieve the lowest cost of production; to make full use of production resources, minimize waste and limitation of the production resources.[2]

句意：所谓优化的生产计划，就是有利于充分利用销售机会，满足市场需求；有利于充分利用盈利机会，并实现生产成本最低化；有利于充分利用生产资源，最大限度地减少生产资源的浪费和限制。

Exercises

1. Summarizing the characteristics of those production process control options.
2. Please describe the steps of the production planning.
3. Explain the definition of production control.
4. Please summarize the tasks and characteristics of each stage of theproduction planning.

Unit 2 The Main Planning and Control Technology of Manufacturing Enterprise

Requirements Forecasting

Forecasting the future has long been a challenge to managers. Fortune tellers, astrologers, priests, and prophets have sought to fulfill people's need to predict the future and reduce its uncertainties. These predictions have not been made just from intellectual curiosity.[1] Knowledge of the future has always promised many kinds of advantages and opportunities.

The forecast is a prediction of future events used for planning purposes. Changing and increasing environmental concerns exert pressure on a firm's capability to generate accurate forecasts. Forecasts are needed to aid in determining what resources are needed, scheduling existing resources, and acquiring additional resources. Accurate forecasts allow schedulers to use capacity efficiently, reduce customer response time, and cut inventories.

Planning Forecasting Requirements

It is probable to do without planning and consequently forecasting. However, this could lead to costly inconveniences that modern societies are unwilling to accept. For instance, it can be envisioned that cooking does not start until one gets hungry, or that a shoe manufacturer does not commence manufacturing ones that have summer styles until there is a demand for such shoes.

Therefore, forecasting must be viewed, as composed of two distinct parts forecasting continuations of established patterns or relationships, and forecasting changes in those pattern and relationships. Fortunately, changes from past patterns or relationships do not happen "every day," which means that "normal" forecasting procedures can be used the great majority of the time. This is advantages because it is easier and more accurate than predicting changes, but it also creates a problem because forecasters tend to forget that basic changes from historical patterns or relationships are possible and in the long run highly likely. Thus people often are caught by surprise when changes occur, and they blame forecasting for the undesirable consequences.

In order to make effective planning, managers must deal with both the continuations of patterns and systematic changes in patterns, and must be able to identify transitions from one to the other. This suggests the need for an expanded role for forecasting. Monitoring and understanding must complement traditional, "extrapolative" forecasting. The end result is final forecasts that might require considerable management adjustments when systematic changes from established patterns and relationships occur. This is where the interface of forecasting and planning becomes critical, the planner or decision-maker must have knowledge of three things as follows:

(1) Extrapolating which plans and strategies are appropriate when the assumption of no change in established historical patterns or relationships is correct.

(2) Monitoring when a systematic change is taking place.

(3) Understanding how such a change will affect the organization.

Designing the Forecasting System

Deciding what to forecast: Although some sort of demand estimate is needed for the individual goods or services produced by a company, forecasting total demand for groups or clusters and then deriving individual product or service forecasts may be easiest. And selecting the correct unit of measurement (e. g., product or service units or machine-hours) for forecasting is also as important as choosing the best method.

Level of aggregation: Few companies make mistakes by more than 5 percent when forecasting total demand for all their products. Even so, errors in forecasts for individual items may be much higher. By clustering several similar products or services in a process called aggregation companies can obtain more accurate forecasts. Many company utilize a two-tier forecasting system, first making forecasts for families of goods or services that have similar demand requirements and common processing, labor, and materials requirements and then deriving forecasts for individual items. This ap-

proach maintains consistency between planning for the final stages of manufacturing (which requires the unit forecasts) and longer-term planning for sales, profit, and capacity (which requires the product family forecasts).

Units of measurement: The most useful forecasts for planning and analyzing operation problems are those based on product or service units, such as customers needing maintenance service or repairs for their cars, rather than dollars. Forecasts of sales revenue are not very helpful because prices often fluctuate. Forecasting the number of units of demand—and then translating these estimates to sales revenue estimates by multiplying them by the price—often is the better method. It accurately forecasting the number of units of demand for a product or service is not possible, forecasting the standard labor or machine hours required of each of the critical resources, based on historical patterns, often is better. For companies producing goods or services to customer order, estimates of labor or machine, hours are important to scheduling and capacity planning.

Choosing the type of forecasting techniques: The forecaster's objective is to develop a useful forecast from the information at hand with the technique appropriate for the different characteristics of demand. This choice sometimes involves a trade-off between forecast accuracy and costs, such as software purchases, the time required to develop a forecast, and personnel training. Two general types of forecasting techniques are used for demand forecasting: qualitative methods and quantitative methods.[2] Qualitative methods include judgment methods, which translate the opinions of managers, expert opinions, consumer surveys, and sales-force estimates into quantitative estimates. Quantitative methods include causal methods and time-series analysis. Causal methods use historical data on independent variables, such as promotional campaigns, economic conditions, and competitors, actions, to predict demand. Time-series analysis is a statistical approach that relies heavily on historical demand data to project the future size of demand and recognizes trends and seasonal patterns.

Master Production Scheduling

The master production schedule (also commonly referred to as the MPS) is effectively the plan that the company has developed for production, staffing, inventory, etc. It has an input of a variety of data, e. g. forecast demand, production costs, inventory costs, etc. and an output of a production plan detailing amounts to be produced, staffing levels, etc. for each of a number of time periods.

Generally speaking, a production plan is complete specification of the amounts of each end item or final product and subassemblies produced the exact timing of the production lot sizes, and the final schedue of completion. The production plan may be broken down into several component parts: ① the master production scheduling (MPS), ② the materials requirements planning (MRP) system, and ③ the detailed job shop schedules. Each of these parts can represent a large and complex subsystem of the entire plan.

At the heart of the production plan are the forecasts of demand for the end items produced over the planning horizon. An end item is the output of the productive system; that is, products shipped out the door. Components are items in intermediate stages of production, and raw materials are resources that enter the system. It is important to bear in mind that raw material, components, and end items are defined in a relative and not an absolute sense. Hence, we may wish to isolate a portion of a company's operation as a productive system. End items associated with one portion of the company may be raw materials for another portion. A single productive system may be the entire manufacturing operation of the firm or only a small part of it.

The master production scheduling (MPS) is a specification of the exact amounts and timing of production of each of the end items in a productive system. The MPS refers to exaggerated item. As such, the inputs for determining the MPS are forecasts for future demand by item rather than by aggregate items. The MPS is then broken down into a detailed schedule of production of each of the components that comprise an end item. The materials requirements planning (MRP) system is the means by which this is accomplished. Finally, the results of the MRP are translated into specific shop floor schedules and requirements of raw materials.

MRP. MRP Ⅱ and ERP

Manufacturing enterprise planning and control technology has become more sophisticated with the development of IT and Internet technology, from the angle of resource management, can be roughly divided into three stages, as shown in Table 5-1[33]

Table 5-1 The Stages of Manufacturing Enterprise Planning and Control Technology

Stage	Background	The Main Solved Problems	The Main Contents and Features
Material Planning	In the early 1960s, production mode was multi varieties and small batch, there existed excess consumption and resource allocation unreasonable phenomenon	Inventory of raw materials, parts production plan, there were serious gap between production and marketing	The material requirements planning based on related demand principle, minimum input and critical path
(Materials + Production Capacity + the Fund plan)	The competitive environment of enterprises deteriorating; enterprises expand the management scope of their own resources, the demanded of manufacturing resource planning to be detail and accurate	The imbalance of production capacity, production have no touch with other business aspects, especially financial control issues	The planning and control of manpower, equipments and more resources (closed-loop MRP), capital control of manufacturing range, which upgrades to manufacturing resource planning (MRP Ⅱ)
The Resources Plan of Enterprise Category	During 10 years after the generation of MRP concepts, the principles, methods and software of enterprise planning and control are mature and perfect, the appearance of new management methods such as JIT, new management ideas and strategies such as CIMS and lean production (LP)	The management of entire supply chain, integration problems throughout the enterprise resources and environment	The range of planning and control extends from manufacturing department to the entire enterprise, principles and methods of resource planning apply to non-manufacturing industry, which upgrades to enterprise resource planning (ERP)

Material requirements planning (MRP) —a computerized information system—was developed specifically to aid companies manage dependent demand inventory and schedule replenishment order.

MRP systems have proven to be beneficial to many companies. In this section, we discuss the nature of dependent demands and identify some of the benefits firms have experienced with these systems.

Dependent Demand

To illustrate the concept of dependent demand, let us consider a Huffy bicycle produced for retail outlets. The bicycle, one of many different types held in inventory at Huffy's plant, has a high-volume demand rate over time. Demand for a final product such as a bicycle is called independent demand because it is influenced only by market conditions and not by demand for any other type of bicycle held in inventory. Huffy must forecast that demand. However, Huffy also keeps many other items in inventory, including handlebars, pedals, frames, and wheel rims, used to make completed bicycles. Each of these items has a dependent demand because the quantity required is a function of the demand for other items held in inventory. For example, the demand for frames, pedals, and rims is dependent on the production of completed bicycles. Operations can calculate the demand for dependent demand items once the bicycle production devils are announced. For example, every bicycle needs two wheel rims, so 1, 000 completed bicycles need 1, 000 × 2 = 2, 000 rims. Statistical forecasting techniques aren't needed for these items.

The bicycle, or any other good manufactured form one or more components, is called a parent. The wheel rim is an example of a component—an item that may go through one or more operations to be transformed into or become part of one or more than one style of bicycle. The parent-component relationship can cause erratic dependent demand patterns for components. Suppose that every time inventory falls to 500 units, an order for 1, 000 more bicycles is placed, as shown in from inventory, along with other components for the finished product; demand for the rim is shown in Figure 5-3. So, even though customer demand for the finished bicycle is continuous and uniform, the production demand for wheel rims is "lumpy": that is, it occurs sporadically, usually in relatively large quantities. Thus, the production decisions for the assembly of bicycle, which account for the coasts of assembling the bicycles and the projected assembly capacities at the time the decisions are made, determine the demand for rims.

Managing dependent demand inventories is complicated because some components may be subject to both dependent and independent demand. For example, operations needs 2, 000 wheel rims for the new bicycles, but the company also sells replacement rims for old bicycles directly to retail outlets. This practice places an independent demand on the inventory of rims. Material requirements planning can be used in complex situations involving components that may have independent demand as well as dependent demand inventories.

Benefits of Material Requirements Planning

For years, many companies tried to manage production and delivery of dependent demand inventories with independent demand systems, but the outcome was seldom satisfactory. However, because it recognizes dependent demands, the MRP system enables businesses to reduce inventory levels, utilize labor and facilities better, and improve customer service. These successes are due to three advantages of material requirements planning.

(1) Statistical forecasting for components with lumpy demand results in large forecasting errors. Compensating for such errors by increasing safety stock is costly, with no guarantee that stock outs can be avoided. MRP calculates the dependent demand of components from the production schedules of their parents, thereby providing a better forecast of component requirements.

(2) MRP systems provide managers with information useful for planning capacities and estimating financial requirements. Production schedules and materials purchases can be translated into capacity requirements and dollar amounts and can be projected in the time periods when they will appear. Planners can use the information on parent item schedules to identify times when needed components may be unavailable because of capacity shortages, supplier delivery delays, and the like.

(3) MRP systems automatically update the dependent demand and inventory replenishment schedules of components when the production schedules of parent items change. The MRP system alerts the planners whenever action is needed on any component.

Manufacturing

The purpose of an MRP system is to reduce the cash needed by a manufacturing organization. This increases the organization's return on investment, directly making the manufacturer a more profitable, attractive investment.

In a classical manufacturing organization, huge amounts of cash are tied up in in-process inventory, parts that must be assembled and then sold. MRP is used with the planning and management to reduce this cash to the minimum.

The basic idea is simple. A sales or marketing group estimates how many products it will sell at a future time. The MRP software backdates using the factory's estimated time to assemble each product. Then, the system explodes the product into lists of parts needed, using the bills of materials developed by Engineering. Department The parts are ordered at times backdated from the assembly dates.

Finally, the cash flow of the above ordering, assembly, ship and payment process is developed. The system provides reports about which parts are needed to ship an order. If a high-value order is waiting for a few cheap parts, the planner can ask for the parts to be flow in by airfreight, and usually this will improve profits and return on investment.

Manufacturing Resource Planning (MRP Ⅱ)

This is a development that seeks to address some of the shortcomings of MRP. It includes all of the elements of MRP; it is based around the Bill of Materials, uses a Master Production Scheduling (MPS) as its starting point and uses the three steps of Explosion, Netting and Offsetting to create the initial schedule.

However MRP Ⅱ includes the following four major developments from MRP:

Feedback

MRP Ⅱ includes feedback from the shop floor on how the work has progressed, to all levels of the schedule so that the next run can be updated on a regular basis. For this reason it is sometimes called 'Closed Loop MRP'.

Resource Scheduling

There is a scheduling capability within the heart of the system that concentrates on the resources, i. e. the plant and equipment required to convert the raw materials into finished goods. For this reason the initials 'MRP' now mean Manufacturing Resources Planning. The advantages of this development are that detailed plans can be put to the shop floor and can be reported on by operation, which offers much tighter control over the plant. Moreover loading by resource means that capacity is taken into account. The difficulty is that capacity is only considered after the MRP schedule has been prepared. It may turn out that insufficient time was allowed within the MRP schedule for the individual operations to be completed.

Batching

Rules batching rules can be incorporated; indeed they have to be if resource scheduling is to take place. Most software packages offer a variety of batching rules. Three of the more important are 'Lot for Lot', 'EBQ' and 'Part Period Cover'. 'Lot for Lot' means batches that match the orders. Therefore if a company is planning to make 10 of Product A followed by 20 of Product B, then the batches throughout the process will match this requirement. If both A and B require two of one certain subassembly then that will be made in quantities of 20 of A and 40 of B. It is the batching implicitly followed in basic MRP.

'EBQ' stands for Economic Batch Quantity. The batch size is calculated by a formula that minimizes the cost through balancing the set up cost against the cost of stock.

'Part Period Cover' means making batches whose size cover a fixed period of demand. A policy of making a week's requirement in one batch is an example.

Software Extension Programs

A number of other software programs are included in the MRP Ⅱ suite. Some of these are further designed to help the scheduling procedure. The most important is Rough Cut Capacity Planning (RCCP), an initial attempt to match the order load to the capacity available, by calculating (using a number of simplifying assumptions) the load per resource. Overloads are identified and orders can be moved to achieve a balance. This has been described as "knocking the mountains (the overloads) into the valleys (periods of under load)".

Other additions are designed to extend the application of the MRP Ⅱ package. For example it may include an option for entering and invoicing sales orders (Sales Order Processing). Another common extension is into stock recording and a third into cost accounting. A full MRP Ⅱ implementation can therefore act as an integrated database for the company.

Data Accuracy

This last development means that the company must put great emphasis on data accuracy. Errors in recording in one part of the system will result in problems for all the users. The suppliers of such systems encourage users to aim for accuracy of between 95% and 98%.

Enterprise Resource Planning (ERP)

ERP originated from America in 1960s, first proposed by Gartner in 1993 as a management philosophy. It is a realization of the scientific management thought by computer, it emphasizes on product development and design, operation control, production planning, purchasing, marketing, sales, inventory (input products, semi-finished products, finished products), financial and personnel and other aspects of the integrated optimization of management, and includes the corresponding module component.

The management of the enterprise involved the logistics, information flow and capital flow, so the general management module of ERP mainly includes three aspects: production control (planning, manufacturing), logistics management (distribution, procurement, inventory management) and financial management (accounting, financial management).

We can understand ERP from the following three levels:

First of all, it is a management idea: it is a set of enterprise management system standards, and is a management idea oriented to supply chain on the basis of MRPII.

Second it is a software product: which puts the ERP management thought as the soul, and comprehensive applies of client/server system, the relational database structure, object-oriented technology, the graphical user interface, the fourth generation language (4GL), network communication and other information achievement. [3]

Third, it is a management system: it is an enterprise resource management system which integrated enterprise management concepts, business processes, basic data, human and material resources, computer hardware and software in one.

ERP is a management information system which comprehensive integrated management three big flows (logistics, capital flow, information flow) of enterprises.

Notes

1. Fortune tellers, astrologers, priests, and prophets have sought to fulfill people's need to predict the future and reduce its uncertainties. These predictions have not been made just from intellectual curiosity.

句意：算命者、占星家、牧师和预言家们为满足人们对未来的不确定性的预知而致力于去预测人们的未来。这些预测不仅仅是对未知的好奇心。

2. Two general types of forecasting techniques are used for demand forecasting: qualitative methods and quantitative methods.

句意：两种常用的基本需求预测技术是定量预测法和定性预测法。

3. Second it is a software product: which puts the ERP management thought as the soul, and comprehensive applies of client/server system, the relational database structure, object-oriented technology, the graphical user interface, the fourth generation language

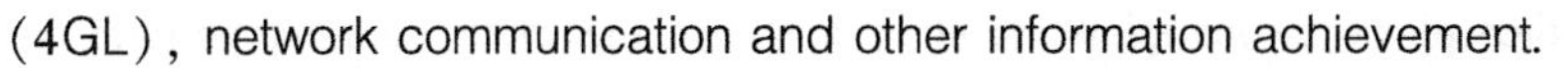

(4GL), network communication and other information achievement.

句意：软件产品：综合应用客户机/服务器体系、关系数据库结构、面向对象技术、图形用户界面、第四代语言（4GL）、网络通信等信息成果，以 ERP 管理思想为灵魂的软件产品。

Exercises

1. Tell the differentia of the MRP, MRPII and ERP.
2. Explain the master production schedule.
3. What are the qualitative method and the main principle of the demand forecasting?
4. Describe the development of the MRPII and the relationship of MRP and MRPII.
5. Summarize the important function of the MPS.

Unit 3 Production Process Control

Production process controlis the analysis, diagnosis and monitoring of operation technology and production process which taken on the production, installation and service process to ensure the production process in a controlled state. Those issues effect on the quality of the products directly or indirectly. The greatest concern in production process control is whether we are doing the right things and doing thing rights.

The objects of production process control include output factors (quantity, quality, cost, delivery, service, brand image, etc.); resources (human, machine, material, information and so on); process elements (goals, plans, processes, rules and regulations, incentives and other); environmental factors (natural environment, work environment, system environment, enterprise culture, etc.)

The Main Types of Pull Production Control System

There currently exist three basic topologies for pull production control system, namely CONWIP or Constant Work in Process, KANBAN, and Base Stock PPCS, and there also is a production control system that has taken into account the circular nature of systems. [1] This type of control system studies has been done regarding the analysis of short-term effects and transient response on a PPCS. A critical concept for short-term analysis is a 'disaster', referring to the event of production falling below a percentile base scenario. The term jamming and jamming margin are introduced as substitutes, which are believed to be better terms, in accord with statistical analysis of dynamic systems.

Dynamic systems are, and their existence is not hindered by a disastrous event. Rather, the disaster is a particular dynamic event. In this context, we will justifies the use of jamming as a better suited term for systems analysis, particularly so because it's referred to a dynamic one.

Kanban Control

This is implemented by circulating cards (kanbans). The machine must have a card before it can start an operation. It can then picks raw materials out of its upstream (or input) buffer, perform the operation, attach the card to the finished part, and put it in the downstream (or output) buffer.

Note the return of a card (kanban) from the output buffer to the machine upstream resembles the use of money. The downstream forwarding of kanbans from a machine does not resemble the use of money (Figure 5-3).

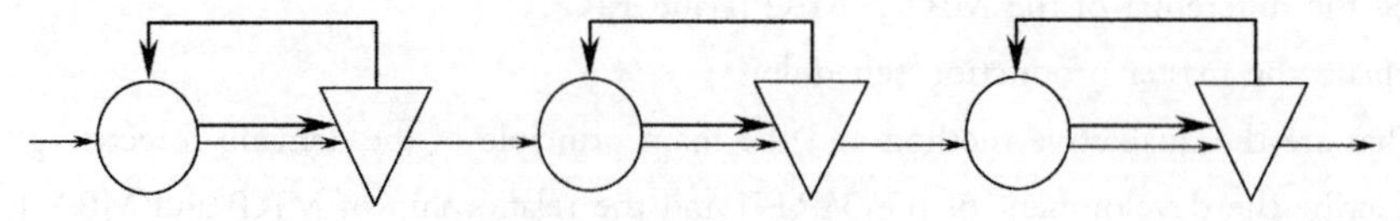

Figure 5-3 Kanban Control

Flow of service shown by thin arrow, flow of kanbans by bold

Constant Work-In-Process (CONWIP) Control

Once the parts are released, they are processed as quickly as possible until they wind up in the last buffer as finished goods. One way to view this is that the system is enveloped in a single kanban cell: Once the consumer removes a part from the finished goods inventory, the first machine in the chain is authorized to load another part (Figure 5-4).

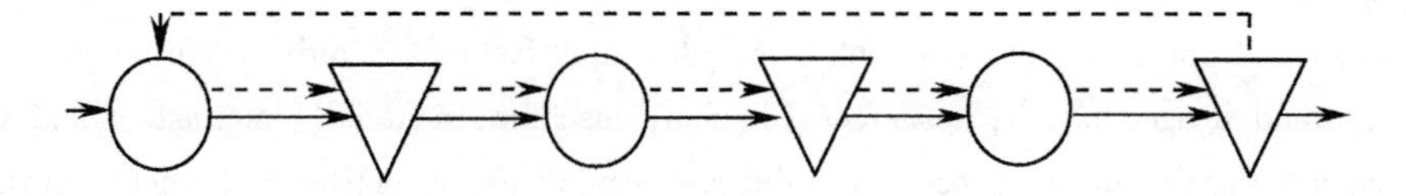

Figure 5-4 C-CONWIP Control

Flow of service shown by solid arrow, and flow of release authorizations by dotted tine

Circular Variable Work-In-Process (C-VARWIP) Control

Visualize a closed productive system where all work is generically named "Service".

A generic service is added value, at times it is evident in an objects attributes, and at times it is reflected without material substance.

Where applicable, a service need not be referenced to a material substance. This aids in defining work as a service rendered in exchange for money where presence of matter is not relevant, but the funneling of generative or degenerative energy is present in the system.

In the C-VARWIP diagram, line of square dots arrows represent the flow of —Pull—cards in the productive system just like it is used in Kanban or CONWIP, the only reference on their directionality is them being transferred conditionally on exchange for a service. What distinguishes these green cards is, they do not flow in company of the service at process termination but they flow locally in reverse direction with synchrony. Line of square dots arrows can represent how traditional money is handled; double arrows represent the flow of a service (Figure 5-5).

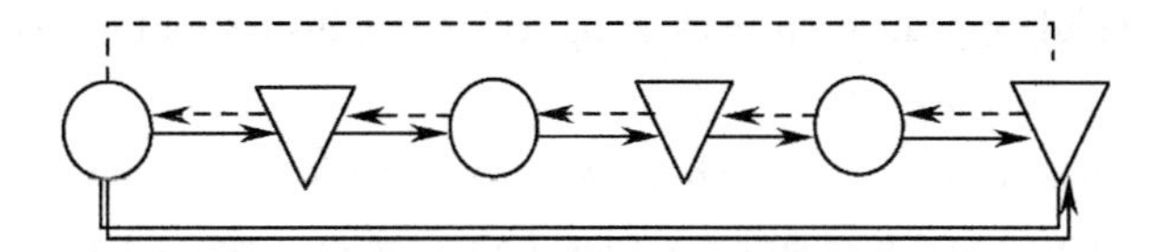

Fig 5-5 C-VARWIP Control

Flow of service shown by solid, and flow of release authorizations by dotted line.

C-VARWIP is a top-down approach to PPCS, and by that nature, it must deal not only with service dynamics but also with money dynamics. Another source of improvement to quality of service at the workstation level is the analysis and re-engineering of the system in its proper circular dimension.

Comparing of Different Production Process Control Approaches

Every manufacturing company is on the road to improvement. While most companies are going the right way on the road, many are unhappy with their slow rate of improvement. As a result, nearly every company is asking serious questions about their processes. "What should we change to next?" American expert has to establish production system simulation model with four kinds of different manufacturing process control approaches to analysis and compare the performance of the four different modes. In this part we will help readers understand the four kinds of production mode through the simulation experiment.

Let's consider a typical production process of 10 available machines of 6 different types. Raw materials enter from the left and flow through the production process towards the right where the products become finished goods. Figure 5-6 shows how each product flows through one of each of the six types of machines. In this linear flow process, there are no assemblies or disassembly processes. Where multiple machines are available of a common type, the product may flow through any one of those machines without a set-up.

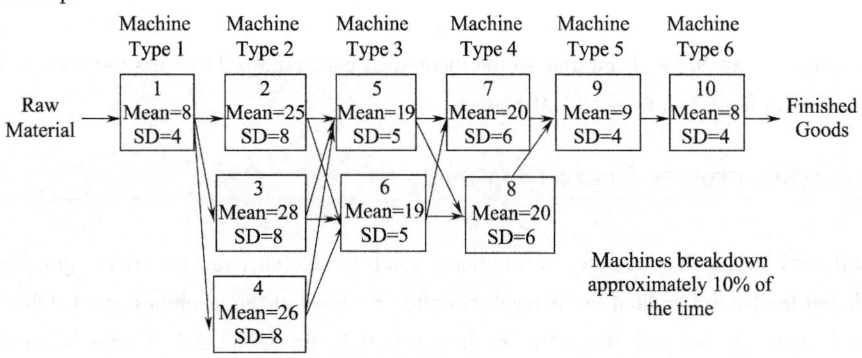

Figure 5-6 Production Simulation Model

There are two different products. Product A accounts for 2/3 of the workload and Product B accounts for 1/3 of the workload. The raw materials arrive according to a weekly schedule. Exactly 120 Product A parts and 90 Product B parts arrive one at a time over 5 each 8 hour shifts (2,400 Minutes). The same schedule repeats each week. Figure 5-7 is a histogram that shows the relative intensity of the arrivals by product over the week.

To add reality to this model, each machine work speed varies according to a normal distribution. The mean (estimated value) and standard deviation (measure of variance) is noted in Figure 5-7. In addition, each machine is subject to random breakdowns. Breakdowns occur every 100 minutes on the average (drawn from a negative exponential distribution) and last about 10 minutes (also drawn from a negative exponential distribution). This means the machines are only available 90% of the time.

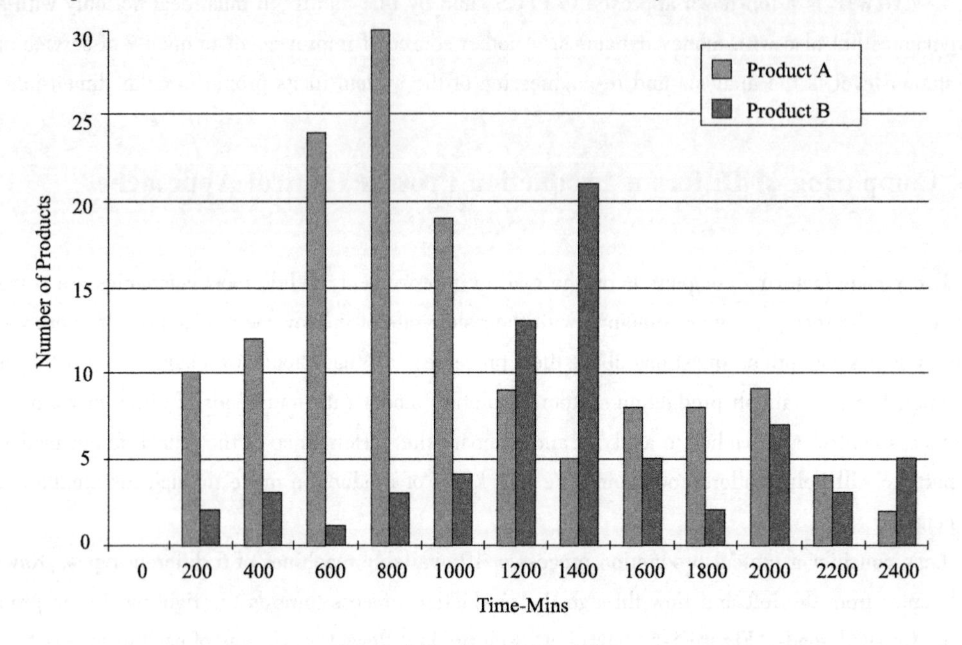

Figure 5-7 Arrival Schedules

Theoretically, this system should be able to produce 216 parts per week. The limiting process is Machine Type 4 which has two machines (7 and 8) processing with a mean of 20 minutes. Machine 7 and 8 produce two parts every 20 minutes or average 10 minutes per part. This is 240 parts per week. These machines operate 90% of the time so the theoretical maximum is 216 parts per week. These simulation models run for 2 1/2 weeks (100 hours).

Production Process Control Options

Traditional Batch Production: In traditional batch production, raw material is grouped together in a batch and the batch is routed to the machines and processed at the machines as a batch. Each part is processed separately but each part waits for the others to be processed and then the batch moves as a whole to the next process. In this model, machine efficiency is emphasized and workflow time is not. This method is common where material handling is a problem. A batch could be a 24-disk cassette, a 120 piece flat pallet or 1000 punched parts in a bin. In the batch simulation model, the batch size is 10. Ten parts all travel together through each process.

Traditional Cell Production: One technique to help manage a complex manufacturing system is to

divide the shared resources into work cells that can be more effectively managed. In this manner a cell of equipment is dedicated to producing one category of product and is not available for other work.[2] In the simulation model, Product A is routed through machines 1→2 or 3→5→7→9→10. Product B is routed through machines 1→4→6→8→9→10. The constraint for Product A is Machine 7 at 20 minutes each. The constraint for Product B is Machine 4 at 26 minutes each.

Just-In-Time: The Just-In-Time production control model is a pull system. Each workstation only produces the quantity of products requested by the customer in subsequent operations. Work control often uses a Kanban card to control the flow of parts through the system. When a customer wants a product, the Kanban card is given to the previous process machine and that machine passes the requested product to the customer. The customer starts work on the product received and the previous process starts work on preparing the next product for the customer. In this simulation model, the demand is from right to left. As a machine on the right needs a product, machines on the left produce the product just in time. Two Just-In-Time simulations are shown here. JIT Kanban of 3 allows 3 products of Work-In-Process between machines.

Theory of Constraints: The Theory of Constraints work control process is based on three elements: The Drum (the speed or capacity of the slowest process), the Buffer (the amount of inventory allowed into the system before the constraint) and the Rope (the releasing mechanism for raw material).[3] In the DBR process, every effort is made to increase the efficiency of the constraint even at the determent of other measures. The inventory level (the buffer) must insure the constraint never runs out of work. The rope mechanism restricts all inventory not needed to protect the constraint from entering the system. Each machine is directed to work as fast as it can when there is work and to sit idle if there is not work.

Lean Manufacturing: Lean manufacturing attempts to maintain low Work-In-Process inventories and avoid excess investments (excess capacity). Lean and be either a push or pull system. Work-In-Process is limited and excess capacity is trimmed away. In this simulation, a push model is used with a maximum of 5 products between machines. And the work process times are slowed to an average of 10 minutes per product for each type of machine (machines 1, 9 and 10 are 10 minutes per product; machines 2, 3 and 4 are 30 minutes per product; machines 5 and 6 are 20 minutes per product). Standard deviations stayed the same.

Agile Manufacturing: The concept of agile manufacturing is flexibility. With agile processes the manufacturing line changes rapidly to changes in demand. If the product demand changes during one part of the week, the resources are realigned to meet the new demand. Agile manufacturing often includes multi-tasking of machines and multi-skilled operators. In this simulation, machine 1 is cross-trained to help machines 2, 3 and 4. Machine 9 is cross trained to help machines 5 and 6. Machine 10 is cross-trained to help machines 7 and 8. The cross trained machines only help the other machines when the queues build up on those machines and the cross trained machine is idle. In each case, the cross-trained machine is not as quick as the machines it is helping.

Performance Measures

Several measurement factors are used to compare the results of these simulations.

WIP—Work-In-Progress. The simulation measures WIP as a rolling average. It uses an exponential smoothed average of WIP over time (alpha value of 0.02).

Flow Time—The Flow Time is the average of the actual flow time from when a product enters the production process until it is completed. Flow Time starts counting when the product raw materials are accepted.

Efficiency—Average Efficiency is a composite measure averaging the utilization of all ten machines.

Produce—The number of finished goods completed over the length of the simulation is reported. Each produced product is sold immediately.

Profit—Profit is measured as the number produced (sales price, raw material price) less operating expenses. The sales price is ¥1,300 for each product. The raw material price is ¥500 per product. Operating costs are ¥3,000 per hour.

ROI—Return on Investment is a measure relating profit to the investment of the firm in fixed assets and in inventory. ROI is profit/ (annualized capital charge plus inventory costs). The annualized capital charge for 100 hours is $50,000. Inventory is charged at raw material cost.

Each model was simulated 20 times for 100 hours. The performance measures are tabulated in Table 5-2.

Several models demonstrated high production capacity: JIT Kanban 3, Lean, and Agile. The models with the lower inventory and quickest response times were marginally better on their rate of return.

Low Work-In-Process levels and short flow times are desirable. It is also desirable to be able to predict delivery times (low variability in flow time). Having enough inventories to keep critical processes operating seems more important than keeping all machines at high levels of efficiency.

Having flexibility to move resources around (as in the agile model) is highly desirable. Overall the DBR control mechanism provides high production levels, low flow times, accurate prediction of flow times and good cost figures.

Table 5-2 Simulation Results

	Traditional Batch		Cell		JIT Kanban of 3		Lean		DBR		Agile	
	Mean	Std Dev	Mean	Mean	Mean	Std Dev	Mean	Std Dev	Mean	Std Dev	Mean	Std Dev
WIP	94	2	24	38	38	2	38	5	19	1	35	2
Flow Time	1,060	19	288	373	373	12	373	36	221	7	395	21
Effic (%)	83	1	73	80	80	2	80	1	76	1	75	1
Produce	436	4	487	472	472	7	472	8	500	3	500	4
Profit	$4,880		$8,960	$7,760	$7,760		$7,760		$10,000		$10,000	
ROI	9%		18%	18%	18%		18%		20%		19%	

Notes

1. There currently exist three basic topologies for pull production control system, namely CONWIP or Constant Work in Process, KANBAN, and Base Stock PPCS, and there also is a production control system that has taken into account the circular nature of systems.

句意：当前对于拉式生产控制系统，主要有三种基本的运作方式，定量在制品或恒量在制品，看板以及基于库存的生产过程控制系统，另外还有一种具有循环特性的控制系统。

2. Traditional Cell Production: One technique to help manage a complex manufacturing system is to divide the shared resources into work cells that can be more effectively managed. In this manner a cell of equipment is dedicated to producing one category of product and is not available for other work.

句意：传统的单元生产是将复杂的制造系统划分成若干单元来进行管理的一种技术，每个单元可对原料进行更有效的加工。这种生产模式中各个单元的设备专门加工某一类产品而不能加工其他种类产品。

3. The theory of constraints work control process is based on three elements: The Drum (the speed or capacity of the slowest process), the Buffer (the amount of inventory allowed into the system before the constraint) and the Rope (the releasing mechanism for raw material).

句意：产品过程控制的约束理论主要由三个要素组成：节拍（最慢工序的速度或生产能力），缓冲（在约束前可能进入系统的库存品数量），以及限度（原料的供应机制）。

Exercises

Questions

1. Comparing three basic topologies for pull production control system and tell the merits of them.

2. Summarizing the characteristics of those production process control options.

3. Point out the differences between the LEAN MANUFACTURING and AGEILE MANUFACTURING.

Translate these sentences into Chinese.

1. Lean manufacturing attempts to maintain low Work-In-Process inventories and avoid excess investments (excess capacity). Lean and be either a push or pull system. Work-In-Process is limited and excess capacity is trimmed away.

2. Production Planning is the process of converting corporate strategy along with market and financial policy into details for the efficient utilization of the production system.

3. Production planning is the principal chin in the whole production management system that is the precondition of plan and control with production system.

4. Forecasting the future has long been a challenge for managers. Fortunetellers, astrologers, priests, and prophets have sought to fulfill people's need to predict the future and reduce its uncertainties. These predictions have not been made just from intellectual curiosity.

5. MRP systems automatically update the dependent demand and inventory replenishment schedules of components when the production schedules of parent items change. The MRP system alerts the planners whenever action is needed on any component.

Chapter 6
Logistics Engineering

Unit 1 Introduction to Logistics

Definition of Logistics

There are many different definitions for logistics in different countries and periods, and they have always changed with the development of society.

The Council of Logistics Management defined the logistics in 1992 as the process of planning, implementing, and controlling the efficient, effective flow and storage of goods, services, and related information from origin to consumption for the purpose of conforming to customer requirements.

And the definition from the MIT Center for Transportation & Logistics is that logistics involves "managing the flow of items, information, cash and ideas through the coordination of supply chain processes and through the strategic addition of place, period and pattern values".

In practice, the terms "logistics" and "supply chain management" are now used interchangeably. Actually logistics and the supply chain are equivalent terms, so the Institute of Logistics (1998) gave the following definition: "Logistics is the time related positioning of resource or the strategic management of the total supply chain. The supply chain is a sequence of events intended to satisfy a customer. It can include procurement, manufacture, distribution, and waste disposal, together with associated transport, storage and information technology.[1]"

In some developed countries, the productivity had risen and led to the saturation of products in the early 1990s, also there were many products which can not be distributed because of the competition. It was difficult to develop the technique as well. People had to find a new way to solve the problems. They intended to expand the market and reduced the cost through the improvement of distribution. It was the initial concept of logistics.

The earliest concept of logistics came from Arch W. Shaw's *Some Problems in Market Distribution* (1915). They called it "Physical Distribution" (PD) at that time. The definition focused mainly on the distribution.

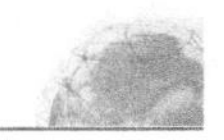

In the late 1980s people had already had a general and deep comprehension about logistics. PD as a definition could not characterize the whole frame of logistics. The logistics included not only the physical distribution but also production logistics, returned logistics, material reuse, etc. Logistics as a suitable definition occurred instead of PD.

Logistics Engineering means the management process of choosing the best scheme under the guidance of theories about system engineering and planning, managing, controlling the system with lowest cost, high efficiency and good customer service for the purpose of improving economy profits of the society and enterprises. [2] In this definition we also integrate the logistics and the flow of information as a system and regard the processes of producing, distribution, and consumption as a whole activity.

Contents and Characteristics

The object of Logistics engineering is to solve the problems in logistics system: the first task is to make system plans and designs with the theory of facility design; and the second one is to manage and control logistics system so as to reduce the cost and improve the efficiency.

Facility Design

Facility design has been often used in industry department such as factory including layout designs, system designs of materials handling, building designs, information system designs, etc.

Logistics Management

1. Logistics System Design

In external logistics system, the design implies the decisions about spots of the networks for materials distribution. But in internal logistics system, the main target is to improve the economy profits of the production system.

2. Transportation/Handling and Storage

It contains:

(1) The research of production optimization.

(2) The research of work station and warehousing storage.

(3) The work-in-process (WIP) products management.

(4) The methods of planning and organizing the handling vehicles.

(5) The organizing methods of information flow and the function in logistics.

3. Handling Facilities, Containers and Package Design

It contains:

(1) The study of warehouse and handling facilities.

(2) The study of handling vehicles and facilities.

In these contents, people often call the first two as the "Soft System", and the last one "Hard System".

The traditional logistics is defined as the simple pattern of materials storage and transportation. The pattern of modern logistics becomes more complex. It can be characterized by:

① Rapid logistics response;

② Serial services;
③ Normative operation;
④ Systematize target and modernized methods;
⑤ Networking organization;
⑥ Marketing business;
⑦ Electronic information.

Functions of Logistics Systems

(1) Members in logistics system should collaborate with the active partners and integrate the serial activities of supply chain to improve the management and strengthen the integrated service capability.

(2) It is helpful for building the rapid response system. The time to prepare and the cost will be reduced. The enterprise will be more competitive.

(3) A good logistics system can reduce the level of organization, work out the personal potentiality of the employees, encourage the team spirit and make sure of maximization of the corporate comprehensive interests. [3]It can also form the active corporate culture oriented by the customer demands and supported by the technological innovation.

Significance of Logistics

The theory of logistics engineering is the study of analyzing, designing, optimizing and controlling the logistics system as a whole. It utilizes the methods of the industrial engineering and system engineering. The study of logistics is of great importance in the production practice.

It can cutback the appropriation of labor, reduce the labor intensity, shorten the production cycle and accelerate the capital turnover. It can also reduce the cost and the circulating capital appropriation, raise the profits and economic efficiency, and improve the product quality and the competitiveness of the enterprise.

Development Trend

When economic globalization, information and networking engulf the entire world, the revolution of the logistics is coming quietly.

1. Logistics Integration

(1) Functional integration-logistics systematized.

(2) Resource and marketing integration-logistics globalization and acquisitions.

2. Logistics Alliances

(1) Vertical logistics alliances—the supply chain.

(2) Horizontal logistics—socialize joint distribution.

(3) External alliances—the third-party logistics.

3. Internet and Information Logistics

(1) Internet logistics—the fusion of the logistics and e-business, procurement and distribution on Internet.

(2) Information logistics—electronic data interchange (EDI).

4. Green Logistics and Virtual Warehousing

Logistics is changing at a rapid and accelerating rate. It is changing for many sets of reasons. The main sets are the pressure for change arising from managerial and technical development within the logistics system itself and the wider economy. Competitive pressures in market are also growing. These, in turn, place pressure on logistics systems.

Notes

1. "Logistics is the time related positioning of resource or the strategic management of the total supply chain. The supply chain is a sequence of events intended to satisfy a customer. It can include procurement, manufacture, distribution, and waste disposal, together with associated transport, storage and information technology."

句意：物流是关于时间的资源配置或者总供应链的战略管理。供应链是满足顾客的事件序列。它可能包括采购、制造、配送、废物处理，以及相伴随的运输、存储和信息技术。

2. Logistics Engineering means the management process of choosing the best scheme under the guidance of theories about system engineering and planning, managing, controlling the system with lowest cost, high efficiency and good customer service for the purpose of improving economy profits of the society and enterprises.

句意：物流工程是指为了改善社会和企业的经济效益，在关于系统工程理论的指导下选择最优的方案，以最低的成本、最高的效率和最好的客户服务规划、管理和控制系统的管理过程。

3. A good logistics system can reduce the level of organization, work out the personal potentiality of the employees, encourage the team spirit and make sure of maximization of the corporate comprehensive interests.

句意：一个好的物流系统可以减少管理的层次，发挥员工的个人潜能，激发团队精神以及确保企业综合利益的最大化。

Exercises

1. What are the main activities included in logistics management?
2. What are the functions of logistics systems?
3. Combining an actual example, discuss the significance of logistics.

Unit 2 Inventory Management

Inventory decisions are high risky and have big impact on supply chain management. Without a proper inventory assortment lost sales and customer dissatisfaction may occur. Likewise, inventory planning is critical to manufacturing. Material or component shortages can shut down a manufacturing line or force modification of a production schedule, which creates added cost and potential finished goods shortages. Just as shortages can disrupt planned marketing and manufacturing operations, inventory overstocks also create operating problems. Overstocks increase cost and reduce profitability as a result of added warehousing, working capital, insurance, taxes, and obsolescence. Management of inventory resources requires an understanding of the principles, cost, impact, and dynamics.

Inventory Risk

Inventory management is risky, and risk varies depending upon a firm's position in the distribution channel. The typical measures of inventory commitment are time duration, depth, and width of commitment.

For a manufacturer, inventory risk is long term. The manufacturer's inventory commitment starts with raw material and component parts, includes work-in-process, and ends with finished goods. In addition, finished goods are often positioned in warehouses in anticipation of customer demand. In some situations, manufacturers are required to consign inventory to customer facilities. In effect, this practice shifts all inventory risk to the manufacturer. Although a manufacturer typically has a narrower product line than a retailer or wholesaler, the manufacturer's inventory commitment is deep and of long duration.

A wholesaler purchases large quantities from manufacturers and sells smaller quantities to retailers. The economic justification of a wholesaler is the capability to provide retail customers with assorted merchandise from different manufacturers in specific quantifies. When products are seasonal, the wholesaler may be required to take an inventory position far in advance of the selling season, thus increasing depth and duration of risk. One of the greatest challenges of wholesaling is product-line expansion to the point where the width of inventory risk approaches that of the retailer while depth and duration of risk remain characteristic of traditional wholesaling.[1] In recent years, retail clientele have also forced a substantial increase in depth and duration by shifting inventory responsibility back to wholesalers.

For a retailer, inventory management is about buying and selling velocity. The retailer purchases a wide variety of products and assumes a substantial risk in the marketing process. Retailer inventory risk can be viewed as wide but not deep. Due to the high cost of store location, retailers place prime emphasis on inventory turnover and direct product profitability. Inventory turnover is a measure of inventory velocity and is calculated as the ratio of annual sales divided by average inventory.

If a business plans to operate at more than one level of the distribution channel, it must be prepared to assume related inventory risk. For example, the food chain that operates a regional warehouse assumes risk related to the wholesaler operation over and above the normal retail operations. To the extent that an enterprise becomes vertically integrated, inventory must be managed at all levels of the supply chain.

Inventory Functionality

From an inventory perspective, the ideal situation would be a response capability to manufacture

products to customer specification. While a zero-inventory manufacturing/distribution system is typically not attainable, it is important to remember that each dollar invested is a trade-off with an alternative use of assets that may provide a better return.

Inventory is a major asset that should provide return for the capital invested. The return on inventory investments is the marginal profit on sales that would not occur without inventory. Accounting experts have long recognized that measuring the true cost and benefits of inventory on the corporate profit-and-loss is difficult. Lack of measurement sophistication makes it difficult to evaluate the trade-offs among service levels, operating efficiencies, and inventory levels. While aggregate inventory levels have been decreased, many enterprises still carry an average inventory that exceeds their basic requirements. This generalization can be understood better through a review of the four prime functions of inventory.

Geographical Specialization

Allow geographical positioning across multiple manufacturing and distributive units of an enterprise. Inventory maintained at different locations and stages of the value-creation process allows specialization.

Decoupling

Allows economy of scale within a single facility and permits each process to operate at maximum efficiency rather than having the speed of the entire process constrained by the slowest.

Supply/Demand Balancing

Accommodate elapsed time between inventory availability (manufacturing, growing, or extraction) and consumption.

Buffering Uncertainty

Accommodates uncertainty related to demand in excess of forecast or unexpected delays in order receipt and order processing on delivery and is typically referred to as safety stock.

These four functions require inventory investment to achieve managerial operating objectives. Given a specific manufacturing/marketing strategy, inventories planned and committed to operations can only be reduced to a level consistent with performing the four inventory functions.[2] All inventories exceeding the minimum level are excess commitments.

At the minimum level, inventory invested to achieve geographical specialization and decoupling can only be modified by changes in facility location and operational processes of the enterprise. The minimum level of inventory required to balance supply and demand depends on the difficult task of estimating seasonal requirements. With accumulated experience over a number of seasonal periods, the inventory required to achieve marginal sales during periods of high demand can be projected fairly well. A seasonal inventory plan can be formulated based upon this experience.

Inventories committed to safety stocks represent the greatest potential for improved logistical performance. These commitments are operational in nature and can be adjusted rapidly in the event of an error or policy change. A variety of techniques are available to assist management in planning safety stock commitments. The focus in the balance of this chapter is on a thorough analysis of safety stock relationships and policy formulation.

Inventory management is a major element of logistical strategy that must be integrated to meet service objectives. While one strategy to achieve a high service level is to increase inventory, other alternative approaches are the use of fast transportation and collaboration with customers and service providers to reduce uncertainty.

Inventory Planning

Key parameters and procedures, namely, when to order, how much to order, and inventory control, guide inventory planning. The when-to-order is determined by the demand and performance average and variationThe how much to order is determined by the order quantity. Inventory control determines the process for monitoring inventory status.

Determining When to Order

The reorder point defines when a replenishment shipment should be initiated. A reorder point can be specified in terms of units or days' supply. This discussion focuses on determining reorder points under conditions of demand and performance cycle certainty.

The basic reorder point formula is

$$R = D \times T$$

where R = Reorder point in units;

D = Average daily demand in units; and

T = Average performance cycle length in days.

To illustrate this calculation, assume demand of 20 units/day and a 10-day performance cycle. In this case,

$$\begin{aligned} R &= D \times T \\ &= 20\text{units/day} \times 10\text{days} \\ &= 200\text{units}. \end{aligned}$$

An alternative form is to define reorder point in terms of days of supply. For the above example, the days of supply reorder point are 10 days.

The use of reorder point formulations implies that the replenishment shipment will arrive as scheduled. When uncertainty exists in either demand or performance cycle length, safety stock is required. When safety stock is necessary to accommodate uncertainty, the reorder point formula is

$$R = D \times T + SS$$

where SS = Safety Stock in units.

Determining How Much to Order

Lot sizing balances inventory carrying cost with cost of ordering. The key to understanding the relationship is to remember that average inventory is equal to one-half the order quantity. Therefore, the greater the order quantity is, the larger the average inventory is and, consequently, the greater the annual carrying cost is. However, the larger the order quantity, the fewer orders required per planning period and, consequently, the lower the total ordering cost. Lot quantity formulations identify the precise quantities at which the annual combined total inventory carrying and ordering cost is lowest for a given sales volume. Figure 6-1 illustrates the basic relationships. The point at which the sum of ordering and carrying cost is minimized represents the lowest total cost. Simply stated, the objectives are to identify the ordering quantity that minimizes the total inventory carrying and ordering cost.

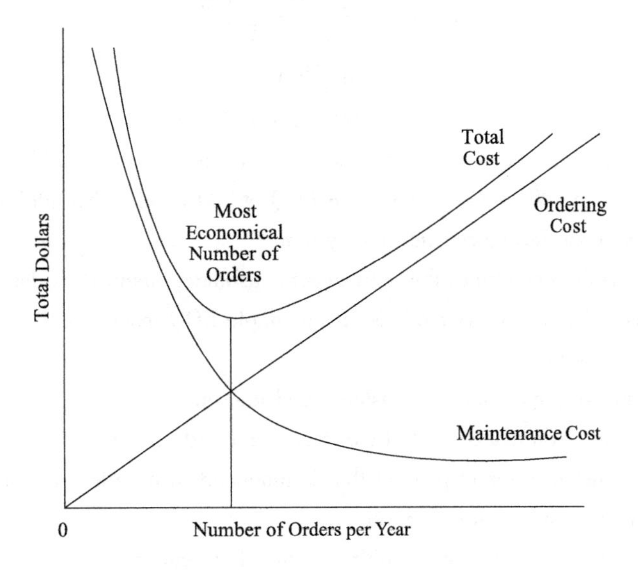

Figure 6-1 Economic Order Quantity

Economic Order Quantity (EOQ)

The EOQ is the replenishment practice that minimizes the combined inventory carrying and ordering cost. Identification of such a quantity assumes that demand and costs are relatively stable throughout the year. Since EOQ is calculated on an individual product basis, the basic formulation does not consider the impact of joint ordering of products.

The most efficient method for calculating EOQ is mathematical. For example a policy dilemma regarding whether to order 100, 200, or 600 units was discussed. The answer can be found by calculating the applicable EOQ for the situation. Table 6-1 contains the necessary information.

Table 6-1 Factors for Determining EOQ

Annual demand volume	2, 400 units
Unit value at cost	$ 5. 00
Inventory carrying cost percent	20% annually
Ordering cost	$ 19. 00 per order

To make the appropriate calculations, the standard formulation for EOQ is

$$\text{EOQ} = \sqrt{\frac{2C_0D}{C_iU}}$$

where EOQ = Economic order quantity;

C_0 = Cost per order;

C_i = Annual inventory carrying cost;

D = Annual sales volume, units; and

U = Cost per unit.

Substituting from Table 6-1:

$$EOQ = \sqrt{\frac{2 \times 19 \times 2400}{0.20 \times 5.00}}$$
$$= \sqrt{91,200}$$
$$= 302\,(\textit{round to } 300)$$

To benefit from the most economical purchase arrangement, orders should be placed in the quantity of 300 units rather than 100, 200, or 600. An EOQ of 300 implies that additional inventory in the form of base stock has been introduced into the system.

While the EOQ model determines the optimal replenishment quantity, it does require some rather stringent assumptions. The major assumptions of the simple EOQ model are:

(1) All demand is satisfied.

(2) Rate of demand is continuous, constant, and known.

(3) Replenishment performance cycle time is constant and known.

(4) There is a constant price of product that is independent of order quantity or time.

(5) There is an infinite planning horizon.

(6) There is no interaction between multiple items of inventory.

(7) No inventory is in transit.

(8) No limit is placed on capital availability.

The constraints imposed by some of these assumptions can be overcome through computational extensions; however, the EOQ concept illustrates the importance of the trade-offs associated with inventory carrying and replenishment ordering cost.

Relationships involving the inventory performance cycle, inventory cost, and economic order formulations are useful for guiding inventory planning. First, the EOQ is found at the point where annualized order cost and inventory carrying cost are equal. Second, average base inventory equals one-half order quantity. Third, the value of the inventory unit, all other things being equal, will have a direct relationship with replenishment order frequency. In effect, the higher the product value, the more frequently it will be ordered.

Inventory Management Policies

Inventory management is the process that implements inventory policies. The reactive or pull inventory approach uses customer demand to pull product through the distribution channel. An alternative philosophy is a planning approach that proactively allocates inventory based on forecasted demand and product philosophy.

Inventory Control

Inventory control is the managerial procedure for implementing an inventory policy. The accountability aspect of control measures units on hand at a specific location and tracks additions and deletions. Accountability and tracking can be performed on a manual or computerized basis.

Inventory control defines how often inventory levels are reviewed to determine when and how much to order. It is performed on either a perpetual or a periodic basis.

Perpetual Review

A perpetual inventory control process reviews inventory status daily to determine inventory replenishment needs. To utilize perpetual review, accurate tracking of all SKUs (Stock Keeping Units) is necessary. Perpetual review is implemented through a reorder point and order quantity.

Periodic Review

Periodic inventory control reviews the inventory status of an item at regular time intervals such as weekly or monthly. For periodic review, the basic recorder point must be adjusted to consider the extended intervals between reviews.

Summary

Inventory typically represents the second largest component of logistics cost next to transportation. The risks associated with holding inventory increase as products move down the supply chain closer to the customer because the potential of having the product in the wrong place or form increases and costs have been incurred to move the product down the channel. [3] Further, the cost of carrying inventory is significantly influenced by the cost of the capital tied up in the inventory. Geographic specialization, decoupling, supply/demand balancing, and buffering uncertainty provide the basic rationale for maintaining inventory. While there is substantial interest in reducing overall supply chain inventory, inventory does add value and can result in lower overall supply chain costs with appropriate trade-offs.

Notes

1. One of the greatest challenges of wholesaling is product-line expansion to the point where the width of inventory risk approaches that of the retailer while depth and duration of risk remain characteristic of traditional wholesaling.

句意：批发业最大的挑战之一就是产品线扩展到一定程度，这时库存风险的宽度与零售商相同，而风险的深度和持续时间仍保持传统批发业的特征。

2. Given a specific manufacturing/marketing strategy, inventories planned and committed to operations can only be reduced to a level consistent with performing the four inventory functions.

句意：对于一个特定的生产或者营销策略，规划的库存和参与运转的库存只能减少到和履行库存的四个基本职能相一致的水平。

3. The risks associated with holding inventory increase as products move down the supply chain closer to the customer because the potential of having the product in the wrong place or form increases and costs have been incurred to move the product down the channel.

句意：随着产品沿着供应链越来越靠近地向消费者移动，与持有库存相关的风险要增加，这是因为产品处于错误的位置或者形式的潜在可能性增长，而且产品在配送渠道中附加了成本。

Exercises

1. Why is it to say that inventory management is necessary to an organization but risky?
2. What are the four basic functions of inventory?

3. When an enterprise makes an inventory planning, what should be done to guarantee its feasibility?

Unit 3 Procurement and Warehousing

Every organization, whether it is a manufacturer, wholesaler, or retailer, buys materials, services, and supplies from outside suppliers to support its operations. Historically, the process of acquiring these needed inputs has been considered somewhat of a nuisance, at least as compared to other activities within the firm. Purchasing was regarded as a clerical or low-level managerial activity charged with responsibility to execute and process orders initiated elsewhere in the organization. [1]The role of purchasing was to obtain the desired resource at the lowest possible purchase price from a supplier. This traditional view of purchasing has changed substantially in the past two decades. The modem focus on supply chain management with its emphasis on relationship between buyers and sellers has elevated purchasing to a higher, strategic-level activity. This strategic role is differentiated from the traditional through the term procurement, although in practice many people use the terms purchasing and procurement interchangeable.

Procurement Perspectives

The evolving focus on procurement as a key capability in organizations has stimulated a new perspective regarding its role in supply chain management. The emphasis has shifted from adversarial, transaction-focused negotiation with suppliers to ensuring that the firm is positioned to implement its manufacturing and marketing strategies with support from its supply base. In particular, considerable focus is placed on ensuring supply, inventory minimization, quality improvement, supplier development, and lowest total cost of ownership.

Procurement Strategies

Effective procurement strategy to support supply chain management concepts requires a much closer working relationship between buyers and sellers than was traditionally practiced. Specially, three strategies have emerged: volume consolidation, supplier operational integration, and value management. Each of these strategies requires an increasing degree of interaction between supply chain partners; thus, they may not be considered as distinct and separate but rather as evolutionary stages of development.

Volume Consolidation

The first step in developing an effective procurement strategy is volume consolidation through reduction in the number of suppliers. At the beginning of the 1980s, many firms faced the reality that they dealt with a large number of suppliers for almost every material or input used throughout the organization. In fact, purchasing literature prior to that time emphasized that multiple sources of supply were the best procurement strategy.

By consolidating volumes with a reduced number of suppliers, procurement is able to leverage its share of a supplier's business. At the very least, it raises the buyer's negotiating strength in relationship to the supplier. More importantly, volume consolidation with a reduced number of suppliers provides a number of advantages for those suppliers. As working relationships with a smaller number of suppliers are developed, those suppliers can, in turn, pass these advantages to the buying organization. The most obvious source of advantage is that by concentrating a larger volume of purchases with a supplier, the supplier can gain greater economies of scale in its own internal processes, partially by being able to spread its fixed costs over a larger volume of output.

Additionally, if a supplier can be assured of a larger volume of purchase, it may be more willing to make investments in capacity or in processes to improve customer service. When a buyer is constantly switching suppliers, no firm has an incentive to make such investments.

It should be noted that volume consolidation does not necessarily mean that a single source of supply is utilized for every, or any, purchased input. It does mean that a substantially smaller number of suppliers are used than was traditionally the case in most organizations. Even when a single source is chosen, it is wise to have a contingency plan in place.

Supplier Operational Integration

The next level of development in procurement strategy emerges as buyers and sellers begin to integrate their processes and activities in an attempt to achieve substantial operational performance improvement in the supply chain. The integration begins to take the form of alliances or partnerships with selected participants in the supply base to reduce the total costs and improve the operating flows between the buyer and the seller.

Value Management

Achieving operational integration with suppliers leads quite naturally to the next level of development in procurement strategy, value management. Value management is an even more intense aspect of supplier integration, going beyond a focus on buyer-seller operations to a more comprehensive relationship. Value engineering reduces complexity and early supplier involvement in product design represents some of the ways that procurement can work with suppliers to reduce further the TCO (Total Cost of Ownership).

Value engineering is a concept that involves closely examining material and component requirements at the early stage of product design to ensure that the lowest total cost inputs are incorporated into that design.[2]

As a firm's new product development process proceeds from idea generation through its various stages to final commercialization, the company's flexibility in making design changes decreases. Design changes can easily be accommodated in the early stages, but by the time prototypes have been developed, a design change is extremely difficult. The expense associated with a design change has the opposite pattern, becoming extremely high after a prototype is developed. The earlier a supplier is involved in the process, then, the more likely an organization is to capitalize on the supplier's knowledge and capabilities.

Purchase Requirement Segmentation

The Pareto effect applies in procurement just as it applies in almost every facet of business activity. In the procurement context, it can be stated simply: a small percentage of the materials, items, and services that are acquired account for a large percentage of the dollars spent. The point is that all procured inputs are not equal; however, many organizations use the same approach and procedures for procuring small items as they do for acquiring their most strategic inputs. One result is that they spend as much in acquiring a large $10,000 order of raw materials as they do for a $100 order of copying paper. Since all purchased inputs are not equal, many firms have begun to pay attention to segmented purchase requirements and prioritizing resources and expertise against those requirements.

It would be a mistake, though, to simply use dollar expenditure as the basis for segmenting requirements. Some inputs are strategic materials; others are not. Some inputs have potential for high impact on the business success; others do not. Some purchases are very complex and high risk; others are not. For example, failure to have seat assemblies delivered to an auto assembly line on time could be catastrophic, while failure to have cleaning supplies might be merely a nuisance.

The key is for the organization to apply the appropriate approach to procurement as needed.

E-Commerce and Procurement

The explosion in technology and information systems is having a major impact on the procurement activity of many major organizations. Much of the actual day-to-day work in procurement has traditionally been accomplished manually with significant amounts of paperwork, resulting in slow processes subject to considerable human error. Applying technology to procurement has considerable potential to speed the process, reduce errors, and lower cost related to acquisition.

Basic Electronic Procurement

Probably the most prevalent use of electronic commerce in procurement is Electronic Data Interchange (EDI). EDI, as the term implies, is simply the electronic transmission of data between a firm and its supplier. This allows two or more companies to obtain and provide much more timely and accurate information. The explosion in EDI usage during the late 1990s was a direct recognition of its benefits, including standardization of data, more accurate information, more timely information, and shortening of lead times with associated reductions in inventories.

At its most basic level, EDI is a major component of integration between buyers and sellers. At least in theory, buyers can communicate quickly, accurately, and interactively with suppliers about requirements, schedules, orders, invoices, and so forth. [3]It provides a tool for transparency between organizations, which is needed to integrate processes in the supply chain.

Another basic application of electronic commerce in procurement has been the development of electronic catalogs. In fact, making information available about products and who can supply them is a natural application for computer technology. Electronic catalogs allow buyers to gain rapid access to product information, specifications, and pricing. When tied to EDI systems, electronic catalogs allow buyers to quickly identify and place orders for needed items. Many companies have developed their own online electronic catalogs and efforts have also been devoted to developing catalogs containing products from many suppliers, which will allow buyers to compare features, specifications, and prices very rapidly. These tools potentially can bring significant savings in procurement, especially for stand-

ard items for which the primary criterion is purchase price.

The Internet and B2B Procurement

The real excitement in procurement related to e-commerce is the development of the Internet as a B2B (business-to-business) tool. Even more so than in the business-to-consumer realm, the Internet and the World Wide Web are expected to have a major impact on how businesses interact with one another. As early as 1996, several major organizations, including General Motors and Wal-Mart, announced that suppliers who were not capable of conducting business via the Internet would be eliminated from consideration. Estimates of the future for B2B e-commerce vary even more wildly than business-to-consumer, but at least one respected authority predicts B2B Internet transactions could reach over $6.7 trillion by 2020.

Warehouse Strategy and Functionality

Warehousing incorporates many different aspects of logistics operations. Due to the interaction, warehousing does not fit the neat classification schemes used when discussing order management, inventory, or transportation. A warehouse is typically viewed as a place to hold or store inventory. However, in contemporary logistical systems, warehouse functionality can be more properly viewed as inventory mixing. This section provides a foundation for understanding the value warehousing contributes to logistics. The objective is to introduce general managerial responsibilities related to warehousing.

While effective logistics systems should not be designed to hold inventory for extended times, there are occasions when inventory storage is justified on the basis of cost and service.

Strategic Warehousing

Storage has always been an important aspect of economic development. In the pre-industrial era, storage was performed by individual households forced to function as self-sufficient economic units. Consumers performed warehousing and accepted the attendant risks.

As transportation capability developed, it became possible to engage in specialization. Product storage shifted from households to retailers, wholesalers, and manufacturers. Warehouses stored inventory in the logistics pipeline, serving to coordinate product supply and consumer demand. Because the value of strategic storage was not well understood, warehouses were often considered "necessary evils" that added cost to the distribution process. The concept that middlemen simply increase cost follows from that belief. The need to deliver product assortments was limited. Labor productivity, materials handling efficiency, and inventory turnover were not major concerns during this early era. Because labor was relatively inexpensive, human resources were used freely. Little consideration was given to efficiency in space utilization, work methods, or materials handling. Despite such shortcomings, these initial warehouses provided a necessary bridge between production and marketing.

After World War Ⅱ, managerial attention shifted toward strategic storage. Management began to question the need for vast warehouse networks. In the distributive industries such as wholesaling and retailing, it was traditionally considered best practice to dedicate a warehouse containing a full assortment of inventory to every sales territory. As forecasting and production scheduling techniques improved, management questioned such risky inventory deployment. Production planning became more dependable as disruptions and time delays during manufacturing decreased. Seasonal production and consumption still required warehousing, but overall need for storage to support stable manufacturing and consumption patterns was reduced.

Changing requirements in retailing more than offset any reduction in warehousing obtained as a result of these manufacturing improvements. Retail stores, faced with the challenge of providing consumers an increasing assortment of products, found it more difficult to maintain purchasing and transportation economics when buying from suppliers. [4]The cost of transporting small shipments made direct ordering prohibitive. This created an opportunity to establish strategically located warehouses to provide timely and economical inventory replenishment for retailers. Progressive wholesalers and integrated retailers developed state-of-the-art warehouse systems to logistically support retail replenishment. Thus, the focus on warehousing shifted from passive storage to strategic assortment.

Improvements in retail warehousing efficiency soon were adopted by manufacturing. For manufacturers, strategic warehousing offered a way to reduce holding or dwell time of materials and parts. Warehousing became integral to Just-In-Time (JIT) and stockless production strategies. While the basic notion of JIT is to reduce work-in-process inventory, such manufacturing strategies need dependable logistics. Achieving such logistical support across geography requires strategically located warehouses. Utilizing centralized parts inventory at a central warehouse reduces the need for inventory at each assembly plant. Products can be purchased and shipped to the strategically located central warehouse, taking advantage of consolidated transportation. At the warehouse, products are sorted, sequenced, and shipped to specific manufacturing plants as needed. Where fully integrated, sortation and sequencing facilities become a vital extension of manufacturing.

On the outbound side of manufacturing, warehouses can be used to create product assortments for customer shipment. The capability to receive mixed product shipments offers customers two specific advantages. First, logistical cost is reduced because an assortment of products can be delivered while taking advantage of consolidated transportation. Second, inventory of slow-moving products can be reduced because of the capability to receive smaller quantities as part of a consolidated shipment. Manufacturers that provide assorted product shipments can achieve a competitive advantage.

An important charge in warehousing is maximum flexibility. Such flexibility can often be achieved through information technology. Technology-based applications have influenced almost every area of warehouse operations and created new and better ways to perform storage and handling. Flexibility is also an essential part of being able to respond to expanding customer demand in terms of product assortments and the way shipments are delivered and presented. Information technology facilitates this flexibility by allowing warehouse operators to quickly react to changing customer requirements.

Warehouse Functionality

Benefits realized from strategic warehousing are classified on the basis of cost and service. No warehouse functionality should be included in a logistical system unless it is fully justified on some combination of cost and service basis. Ideally a warehouse will simultaneously provide economic and service benefits.

Economic Benefits

Economic benefits of warehousing occur when overall logistics costs are reduced. For example, if

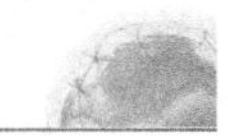

adding a warehouse in a logistical system reduces overall transportation cost by an amount greater than required investment and operational cost, then total cost will be reduced. When total cost reductions are achievable, the warehouse is economically justified.

Service Benefits

Warehouse service can provide benefits through enhanced revenue generation. When a warehouse is primarily justified on service, the supporting rationale is that sales can be increased, in part, by such logistical performance. It is typically difficult to quantify service return-on-investment because it's difficult to measure. For example, establishing a warehouse to service a specific market may increase cost but should also increase market sales, revenue, and potentially gross margin.

Warehouse Operations

Once a warehouse mission is determined, managerial attention focuses on establishing the operation. A typical warehouse contains materials, parts, and finished goods on the move. Warehouse operations consist of break-bulk, storage, and assembly procedures. The objective is to efficiently receive inventory, possibly store it until required by the market, assemble it into complete orders, and initiate movement to customers. This emphasis on product flow renders a modem warehouse as a mixing facility. As such, a great deal of managerial attention concerns how to perform storage to facilitate efficient materials handling.

Handling

The first consideration focuses on movement continuity and scale economies throughout the warehouse. Movement continuity means that it is better for a material handler with a piece of handling equipment to perform longer moves than to undertake a number of short handlings to accomplish the same overall move. Exchanging the product between handlers or moving it from one piece of equipment to another wastes time and increases the potential for product damage. Thus, as a general rule, longer warehouse movements are preferred. Goods, once in motion, should be continuously moved until arrival at their final destination.

Scale economies justify moving the largest quantities or load possible. Instead of moving individual cases, handling procedures should be designed to move cases grouped on pallets, or containers. The overall objective of materials handling is to eventually sort inbound shipments into unique customer assortments.

Storage

The second consideration is that warehouse utilization should position products based upon individual characteristics. The most important product variables to consider in a storage plan are product volume, weight, and storage requirements.

Product volume or velocity is the major factor driving warehouse layout. High-volume product should be positioned in the warehouse to minimize movement distance. For example, high-velocity products should be positioned near doors, primary aisles, and at lower levels in storage racks. Such positioning minimizes warehouse handling and reduces the need for frequent lifting. Conversely, products with low volume should be assigned locations more distant from primary aisles or higher up in storage racks.

Similarly, the storage plan should take into consideration product weight and special characteristics. Relatively heavy items should be assigned storage locations low to the ground to minimize lifting. Bulky or low-density product requires cubic space. Floor space along outside walls is ideal for such items. On the other hand, smaller items may require storage shelves, bins, or drawers. The integrated storage plan must consider individual product characteristics.

Warehouse Planning

Initial decisions related to warehousing are planning based. The basic concept that warehouses provide as an enclosure for material storage and handling requires detailed analysis before the size, type, and shape of the facility can be determined. This section reviews planning issues that establish the character of the warehouse, which in turn determines attainable handling efficiency.

Site Selection

The first task is to identify both the general and then the specific warehouse location. The general area concerns the broad geography where an active warehouse makes sense from a service, economic, and strategic perspective.

Once the combinations of broad areas are determined, a specific building site must be identified. Typical areas in a community for locating warehouses are the commercial zone, outlying areas served primarily by motor truck only, and the central or downtown area.

Design

Warehouse design must consider product movement characteristics. Three factors to be determined during the design process are the number of doors to include in the facility, a cube utilization plan, and product flow.

The ideal warehouse design is a one-floor building that eliminates the need to move product vertically. The use of vertical handling devices, such as elevators and conveyors, to move product from one floor to the next requires time, energy, and typically creates handling bottlenecks. So, while it is not always possible, particularly in central business districts where land is restricted or expensive, as a general rule warehouses should be designed as one-floor operations to facilitate materials handling.

Warehouse design should maximize cubic utilization. Maximum effective warehouse height is limited by the safe lifting capabilities of materials handling equipment, such as lift trucks, rack design, and fire safety regulations imposed by sprinkler systems.[5]

Warehouse design should facilitate continuous straight product flow through the building. This is true whether the product is moving into storage or is being cross-docked. In general, this means that product should be received at one end of a building, stored as necessary in the middle, and shipped from the other end.

Product-Mix Analysis

Another independent area of quantitative analysis is detailed study of products to be distributed through the warehouse. The design and operation of a warehouse are related directly to the product mix. Each product should be analyzed in terms of annual sales, demand, weight, cube, and packa-

ging. It is also desirable to determine the total size, cube, and weight of the average order to be processed through the warehouse. These data provide necessary information for determining warehouse space, design and layout, materials handling equipment, operating procedures, and controls.

Future Expansion

Because warehouses are increasingly important in contemporary logistical networks, their future expansion should be considered during the initial planning phase. Well-managed organizations often establish 5- to 10-year expansion plans. Potential expansion may justify purchase or option of a site three to five times larger than required to support initial construction.

Building design should accommodate future expansion without seriously affecting ongoing operations. Some walls may be constructed of semi permanent materials to allow quick removal. Floor areas, designed to support heavy movements, can be extended during initial construction to facilitate expansion.

Materials Handling Considerations

A materials handling system is the basic driver of warehouse design. As noted previously, product movement and assortment are the main functions of a warehouse. Consequently, the warehouse is viewed as a structure designed to facilitate efficient product flow. It is important to stress that the materials handling system must be selected early in the warehouse development process.

Layout

The layout or storage plan of a warehouse should be planned to facilitate product flow. The layout and the material handling system are integral. In addition, special attention must be given to location, number, and design of receiving and loading docks.

It is difficult to generalize warehouse layouts since they are usually customized to accommodate specific handling requirements. If pallets are utilized, an early step is to determine the appropriate size.

Analysis of product cases and stacking patterns will determine the size of pallet best suited to the operation. Regardless of the size finally selected, management should adopt one pallet size for the overall warehouse.

The second step in planning warehouse layout involves pallet positioning. The most common practice in positioning pallets is 90 degree, or square, placement. Square positioning is widely used because of layout ease.

Finally, the handling equipment must be integrated to finalize layout. The path and tempo of product flow depend upon the materials handling system.

Sizing

Several techniques are available to help estimate warehouse size. Each method begins with a projection of the total volume expected to move through the warehouse during a given period. The projection is used to estimate base and safety stocks for each product to be stocked in the warehouse. Some techniques consider both normal and peak utilization rates. Failure to consider utilization rates can result in overbuilding, with corresponding increase in cost. It is important to note, however, that a major complaint of warehouse managers is underestimation of warehouse size requirements. A good rule of thumb is to allow 10 percent additional space to account for increased volume, new products, and new business opportunities.

Notes

1. Purchasing was regarded as a clerical or low-level managerial activity charged with responsibility to execute and process orders initiated elsewhere in the organization.

句意：采购被认为是一种文书性的或者低层次的管理行为，所承担的责任是执行或者运行企业中其他地方发出的指令。

2. Value engineering is a concept that involves closely examining material and component requirements at the early stage of product design to ensure that the lowest total cost inputs are incorporated into that design.

句意：价值工程是指在产品设计的最早阶段详细检查物资和元件需求的一个概念，它保证设计拥有最低的总成本投入。

3. At its most basic level, EDI is a major component of integration between buyers and sellers. At least in theory, buyers can communicate quickly, accurately, and interactively with suppliers about requirements, schedules, orders, invoices, and so forth.

句意：在最基本的层面上，电子数据交换是买卖双方结合的最基本元件。至少在理论上，买方可以快速地、准确地、交互地和供应者之间就需求、规划、定购、发票等进行交流。

4. Retail stores, faced with the challenge of providing consumers an increasing assortment of products, found it more difficult to maintain purchasing and transportation economics when buying from suppliers.

句意：面临为消费者提供越来越多种类产品的挑战，零售商店发现从供应者购买时维持购买和运输的经济性更为困难。

5. Warehouse design should maximize cubic utilization. Maximum effective warehouse height is limited by the safe lifting capabilities of materials handling equipment, such as lift trucks, rack design, and fire safety regulations imposed by sprinkler systems.

句意：仓库设计应最大限度地利用立方结构，最有效的仓库高度受物资搬运设备，如吊车的安全提升能力、机架的设计和自动喷水灭火系统的防火安全规则的限制。

Exercises

1. What are the procurement strategies?
2. What is the EOQ? Discuss the major assumptions of the simple EOQ model.
3. Discuss the basic principles to guide the selection of materials handling processes and technologies.

Unit 4 Packaging and Materials Handling

Within a warehouse and while being transported throughout a logistics system, the package serves to identify and protect product. The package, containing a product, is the entity that must be moved by a firm's materials handling system. For this reason we will jointly discuss packaging and materials

handling as integral parts of warehousing and a firm's logistical system.

To facilitate handling efficiency, products in the form of cans, bottles, or boxes are typically combined into larger units. This larger unit, typically called the master carton, provides two important features. First, it serves to protect the product during the logistical process. Second, the master carton facilitates ease of handling, by creating one large package rather than a multitude of small, individual products. [1] For efficient handling and transport, master cartons are typically consolidated into larger unit loads. The most common units for master carton consolidation are pallets, slip sheets, and various types of containers.

Package Functionality and Package Operations

Packaging is typically viewed as being either consumer, focused primarily on marketing, or industrial, focused on logistics. The primary concern of logistics operations is industrial package design. Individual products or parts are typically grouped into cartons, bags, bins, or barrels for handling efficiency. Containers used to group individual products are called master cartons. When master cartons are grouped into larger units for handling, the combination is referred to as containerization or unitization.

The master carton and the unitized load become basic handling units for logistical operations. The weight, cube, and damage potential of the master carton determines transportation and materials handling requirements. If the package is not designed for efficient logistical processing, overall system performance suffers.

A prime objective in logistics is to design operations to handle a limited assortment of standardization facilitates materials handling and transportation.

Of course, logistical considerations cannot fully dominate packaging design. The ideal package for materials handling and transportation would be a perfect cube having equal length, depth, and width while achieving maximum possible density. Seldom will such a package exist. The important point is that logistical requirements should be evaluated along with manufacturing, marketing, and product design considerations when finalizing master carton selection.

Another logistical packaging concern is the degree of protection required to cope with the anticipated environments. Package design and material must combine to achieve the desired level of protection without incurring the expense of overprotection. [2] It is possible to design a package that has the correct material content but does not provide the necessary protection. Arriving at a satisfactory packaging solution involves defining the degree of allowable damage in terms of expected overall conditions and then isolating a combination of design and materials capable of meeting those specifications. For package design, there are two key principles. First, the cost of absolute protection will, in most cases, be prohibitive. Second, package construction is properly a blend of design and material.

A final logistics packaging concern is the relationship between the master carton size, retail replenishment quantity, and retail display quantity. From a materials handling perspective, master cartons should be standardized and reasonably large to minimize the number of units handled in the warehouse. For ease of warehouse handling, it is desirable to have retailers purchase in master carton quantities. However, for a slow-moving product, a master carton could contain a substantial overstock for an item that sells only one unit per week but is packed in a case containing 48. Finally, in order to minimize labor, retailers often place trays from master cartons on the retail shelf so that each unit does not have to be unloaded and placed individually. Master cartons or trays meeting retail requirements for shelf space are preferred.

The determination of final package design requires a great deal of testing to assure that both marketing and logistics concern are satisfied. Such tests can be conducted in a laboratory or on an experimental basis. While the marketing aspects are generally the focus of consumer research, logistics packaging research has not been as formalized. During the past decade the process of package design and material selection has become far more scientific. Laboratory analysis offers a reliable way to evaluate package design as a result of advancements in testing equipment and measurement techniques. Instrumented recording equipment is available to measure shock severity and characteristics while a package is in transit. To a large degree, care in design has been further encouraged by increased federal regulation regarding hazardous materials.

Materials Handling

Investments in materials handling technology and equipment offer the potential for substantially improved logistics productivity. Materials handling processes and technologies impact productivity by influencing personnel, space, and capital equipment requirements. Material handling is a key logistics activity that can't be overlooked.

Basic Handling Considerations

Logistical materials handling is concentrated in and around the warehouse. A basic difference exists in the handling of bulk materials and master cartons. Bulk handling is a situation where protective packaging at the master canon level is unnecessary. However, specialized handling equipment is required for bulk unloading, such as for solids, fluids, pellets, or gaseous materials. Bulk handling of such material is generally completed using pipelines or conveyors. The following discussion focuses on the non-bulk handling using master canons.

There are several basic principles to guide the selection of materials handling processes and technologies. The principles summarized in Table 6-2 offer an initial foundation for evaluating materials handling alternatives.

Table 6-2 Principles of Materials Handling

Equipment for handling and storage should be as standardized as possible.
When in motion, the system should be designed to provide maximum continuous product flow.
Investment should be in handling rather than stationary equipment.
Handling equipment should be utilized to the maximum extent possible.
In handling equipment selection the ratio of dead weight to payload should be minimized.
Whenever practical, gravity flow should be incorporated in system design.

Handling systems can be classified as mechanized, semi-automated, automated, and information-

directed. A combination of labor and handling equipment is utilized in mechanized systems to facilitate receiving, processing, and/or shipping. Generally, labor constitutes a high percentage of overall cost in mechanized handling. Automated systems, in contrast, attempt to minimize labor as much as possible by substituting equipment capital investment. When a combination of mechanical and automated systems is used to handle material, the system is referred to as semi-automated. An information-directed system applies computerization to sequence mechanized handling equipment and direct work effort. Mechanized handling systems are most common, but the use of semi-automated and automated systems is increasing. The main drawback to automated handling is lack of flexibility. One factor contributing to low logistical productivity is that information-directed handling has yet to achieve its full potential. This situation is predicted to dramatically change during the first decade of the 21st century.

Notes

1. This larger unit, typically called the master carton, provides two important features. First, it serves to protect the product during the logistical process. Second, the master carton facilitates ease of handling, by creating one large package rather than a multitude of small, individual products.

句意：这种较大的单元，一般被称为主箱，提供两种重要的特性。一是在物流过程中保护产品，二是主箱构建了一个大包装，而不是大量的小而零散的产品，从而便利了搬运。

2. Package design and material must combine to achieve the desired level of protection without incurring the expense of overprotection.

句意：包装的设计和材料的选择应一并满足预期的保护标准，且不发生过保护而引起费用。

Exercises

1. When designing the package of products, what aspects should the designer consider?
2. What are the principles of materials handling?

Unit 5 Supply Chain Management

Brief introduction to supply chain management

What began during the last decade of the 20th century and will continue to unfold well into the 21st century is what historians characterized as the dawning of the information or digital age. In the age of electronic commerce, the reality of B2B connectivity has made possible a new order of business relationship called supply chain management.[1] Managers are increasingly questioning traditional distribution, manufacturing, and purchasing practices. In this new order of affairs, products can be manufactured to exact specifications and rapidly delivered to customers at locations throughout the globe. Logistical systems exist that have the capability to deliver products at exact times. Customer order and delivery of a product can be performed in hours. The frequent occurrence of service failures that characterized the past is increasingly being replaced by a growing managerial commitment to zero defect or what is commonly called the Six-sigma performance. Perfect orders—delivering the desired assortment and quantity of products to the correct location on time, damage-free, and correctly invoiced—once the exception, are now becoming the expectation. Perhaps most important is the fact that such high-level performance is being achieved at lower total cost and with the commitment of fewer financial resources than characteristic of the past.

In this section, the supply chain management business model is introduced as a growing strategic posture of contemporary firms. The supply chain is positioned as the strategic framework within which logistical requirements are identified and related operations must be managed.

Supply Chain Concept

Supply chain (sometimes called the value chain or demand chain) management consists of firms collaborating to leverage strategic positioning and to improve operating efficiency. For each firm involved, the supply chain relationship reflects strategic choice. A supply chain strategy is a channel arrangement based on acknowledged dependency and relationship management. Supply chain operations require managerial processes that span across functional areas within individual firms and link trading partners and customers across organizational boundaries.

At first blush, supply chain management may appear to be a vague concept. A great deal has been written on the subject without much concern for basic definition, structure, or common vocabulary. Confusion exists concerning the appropriate scope of what constitutes a supply chain, to what extent it involves integration with other companies as contrasted to internal operations, and how it is implemented in terms of competitive practices. For most managers, the supply chain concept has intrinsic appeal because it visions new business arrangements offering the potential to improve customer service. The concept also implies a highly efficient and effective network of business linkage that can serve to improve efficiency by eliminating duplicate and nonproductive work.

The supply chain activities encompass all associated with the flow and transformation of goods, the flow of information from the raw materials supplier to the end user, as well as the reverse flow of materials and information in the supply chain.[2]

Supply chain management is the integration of these activities through improved supply chain relationships, to achieve a sustainable competitive advantage.

In this definition, the supply chain includes the management of information systems, sourcing and procurement, production scheduling, order processing, inventory management, warehousing, customer service, and after-market disposition of packaging and materials. The supplier network consists of all organizations that provide inputs, either directly or indirectly, to the focal firm. For example, an automotive company's supplier network includes the thousands of firms that provide items ranging from raw materials such as steel and plastics, to complex assemblies and subassemblies. A given material may pass through multiple processes within multiple suppliers and divisions before being assembled into a vehicle.

The beginning of a supply chain inevitably can be traced back to "Mother earth"; that is, the ultimate original source of all materials that flow through the chain (e. g., iron ore, coal, petroleum, wood, etc.). Supply chains are essentially a series of linked suppliers and customers; every customer is in turn a supplier to the next downstream organization until a finished product reaches the ultimate end user.

A Generalized Supply Chain Model

The general concept of an integrated supply chain is typically illustrated by a line diagram that links participating firms into a coordinated competitive unit. Figure 6-2 illustrates a generalized model adapted from the supply chain management program at Michigan State University.

The context of an integrated supply chain is multi-firm relationship management within a framework characterized by capacity limitations, information, core competencies, capital, and human resource constraints. Within this context, supply chain structure and strategy results from efforts to operationally link an enterprise with customers as well as the supporting distributive and supplier networks to gain competitive advantage. Business operations are therefore integrated from initial material purchase to delivery of products and services to end customers.

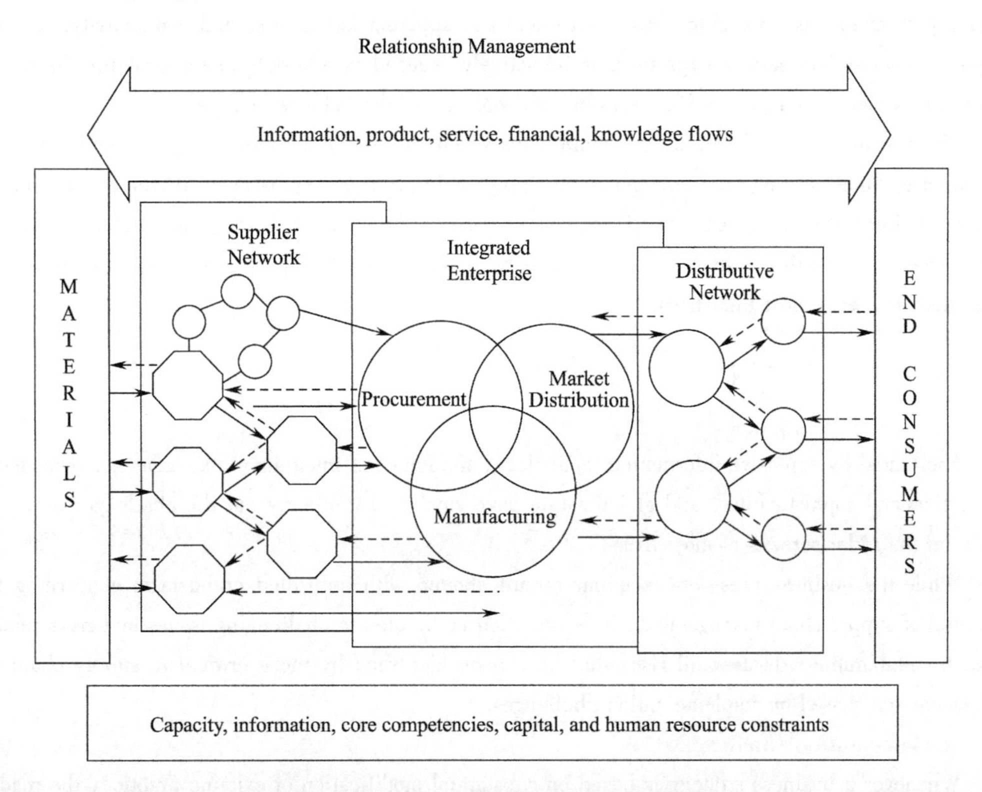

Figure 6-2 A Generalized Supply Chain Model

Value results from the synergy among firms comprising the supply chain with respect to five critical flows: information, product, service, financial, and knowledge (see the bidirectional arrow at the top of Figure 6-2). Logistics is the primary conduit of product and service flow within a supply chain arrangement. Each firm engaged in a supply chain is involved in performing logistics. Such logistical activity may or may not be integrated within that firm and within overall supply chain performance.

The generalized supply chain arrangement illustrated in Figure 6-2 logically and logistically links a firm and its distributive and supplier network to end customers. The message conveyed in the figure is that the integrated value-creation process must be managed from material procurement to end-customer product/service delivery.

The integrated supply chain perspective shifts traditional channel arrangements from loosely linked groups of independent businesses that buy and sell inventory to each other toward a managerially coordinated initiative to increase market impact, overall efficiency, continuous improvement, and competitiveness.[3] In practice, many complexities serve to cloud the simplicity of illustrating supply chains as directional line diagrams. For example, many individual firms simultaneously participate in multiple and competitive supply chains. To the degree that a supply chain becomes the basic unit of competition, firms participating in multiple arrangements may confront loyalty issues related to confidentially and potential conflict of interest.

Another factor that serves to add complexity to understanding supply chain structure is the high degree of mobility and change observable in typical arrangements. It's interesting to observe the fluidity of supply chains as firms enter and exit without any apparent loss of essential connectivity. For example, a firm and/or service supplier may be actively engaged in a supply chain structure during selected times, such as a peak selling season, and not during the balance of a year.

The overarching enabler of supply chain management is information technology. In addition to information technology, the rapid emergence of supply chain arrangements is being driven by four related forces: ① integrative management; ② responsiveness; ③ financial sophistication; and ④ globalization. These forces will continue, for the foreseeable future, to drive supply chain structure and strategy initiatives across most industries.

Issues in Supply Chain Management

Facilitated by explosive information technology, the forces of integrative management, responsiveness, financial sophistication, and globalization have combined to clearly put the challenges of supply chain on the radar screens of most firms.

While the business press and seminar circuit abound with unbridled enthusiasm concerning the potential of supply chain management, little attention is directed to challenging issues and risks related to such collaboration. Issues and risks that have been identified by those critical of supply chain arrangements are based on implementation challenges.

Implementation Challenges

Whenever a business strategy is based on substantial modification of existing practice, the road to implementation is difficult. As noted earlier, the potential of supply chain management is predicated on the ability to modify traditional functional practice to focus on integrated process performance. Such changed behavior requires new practices related to internal integration as well as direction of operations

across the supply chain. To make integrated supply chain practice a reality, at least four operational challenges must be resolved.

Leadership

For a supply chain to achieve perceived benefits for participating firms, it must function as a managed process. Such integrative management requires leadership. Thus, questions regarding supply chain leadership will surface very early in the development of a collaborative arrangement. At the root of most leadership issues are power and risk.

Power determines which firm involved in potential supply chain collaboration will perform the leadership role. Equally important is the willingness of other members of a potential supply chain arrangement to accept a specific firm as the collaborative leader. A supply chain seeking to link manufacturers offering nationally branded consumer merchandise into a supply chain arrangement with a large mass merchandiser that has significant consumer store loyalty can represent substantial power conflict.

Risk issues related to supply chain involvement essentially center on who has the most to gain or lose from the collaboration. Clearly a trucking firm that provides transportation services within a supply chain has far less at stake than either the manufacturer or the mass merchant discussed above. Generally, risk drives commitment to the collaborative arrangement and therefore plays a significant role in determining leadership.

The issue of which firm leads and the willingness of other firms to collaborate under the guidance of such leadership rest at the heart of making the supply chain ideal work.

Loyalty and Confidentiality

In almost every observable situation, firms that participate in a specific supply chain are also simultaneously engaged in other similar arrangements. Some supply chain engagements may be sufficiently different so as not to raise issues of confidentiality.

However, the more common situation is for firms to be engaged as members of supply chain that are direct competitors. From within this maze of competitive interactions, collaborative initiatives must be launched, nurtured, and sustained if the potential of integrated supply chains is to be realized. Firms that simultaneously engage in supply chains that are competitive must develop programs to foster loyalty and maintain confidentiality.

The issues of how to maintain focused loyalty and confidentiality in organizations that simultaneously participate in competitive supply chains are of critical importance. Breaches in confidentiality can have major legal and long-term business consequences. Loyalty quickly comes into question during periods of short supply or otherwise threatened operations. To achieve the benefits of cross-organizational collaboration, these issues must be managed and prospective damage must be controlled.

Measurement

Unlike an individual business, supply chains do not have conventional measurement devices. Whereas an individual business has an income statement and balance sheet constructed in compliance to uniform accounting principles, no such universal documents or procedures exist to measure supply chain performance. The question of supply chain performance is further complicated by the fact that process improvements benefiting overall supply chain performance may reduce costs of one firm while increasing selected costs of other participating firms.

It is clear that the measurement of supply chain operations requires a unique set of metrics that identifies and shares performance and cost information between participating members. The union of multiple firms into a synchronized supply chain initiative requires measures that reflect the collective synthesis while isolating and identifying individual contribution. Likewise, it would be ideal to have supply chain benchmarks to proliferate collective best practices.

The rapid emergence of the supply chain format has helped identify measurement challenges. However, the development of meaningful metrics remains in its infancy.

Risk/Reward Sharing

The ultimate challenge is the equitable distribution of rewards and risks resulting from supply chain collaboration. To illustrate, assume a business situation wherein the leadership of a major manufacturer in collaboration with material suppliers and distributive organizations results in a superior product reaching market at improved profitability. The described scenario is the icon of supply chain success. The product is better than competitors' and is distributed on a more profitable basis. This scenario implies that waste, nonproductive effort, duplication, and unwanted redundancy across the supply chain have been reduced to a minimum while the product and its logistical presentation have reached new heights of achievement. The challenge in success or failure is how to share the benefits or risks.

In traditional practice, the method by which risk and reward are shared is the transfer price. Transfer pricing, guided by market forces, works in transactional driven business relationships. However, supply chain engagements require a higher level of collaboration involving risk and reward sharing. In other words, if the process innovation is successful, the collaborating firms must share benefits. Conversely, if the innovation fails, risks must be appropriately absorbed. While easy to frame in theory, risk and reward sharing arrangements prove extremely difficult to implement in practice.

Clearly without appropriate metrics, it is impossible to share risk or rewards. However, even with the metrics in place, appropriate allocation requires careful preplanning and assessment for sharing programs to work.

Limited Success

The preceding discussion raises some practical limitations regarding the reality of supply chains. The concept of collaborating for success is full of vitality. The mechanics of how to make such complex relationships work on a day-to-day basis are not well understood. Successful supply chain arrangements need to be driven by a well-defined and jointly endorsed set of collaborative principles. Such implementation principles need to prescribe leadership, loyalty and confidentiality measurements, and sharing guidelines and agreements.

In today's world, most so-called supply chains do not enjoy the assumptive and policy framework essential for long-term success. In recent research completed by Michigan State University, it was determined that fewer than one out of five firms engaged in collaborative arrangements had developed and approved policies to guide their managers in the structure and conduct of such arrangements. No firms reported or were willing to share cross-organizational collaborative agreements that extended beyond traditional performance contracts. Some such contracts did contain performance incentives and risk ab-

sorbing agreements; however, such agreements were not as much cooperative arrangements as they were statements of performance expectations.

Summary

The development of greater integrated management skill is critical to continued productivity improvement. Such integrative management must focus on quality improvement at both functional and process levels. In terms of functions, critical work must be performed to the highest degree of efficiency. Processes that create value occur both within individual firms and between firms linked together in collaborative supply chains. Each type of process must be continuously improved.

The supply chain collaborations must be viewed as highly dynamic. Such collaborations are attractive because they offer new horizons for gaining market positioning and operating efficiency. Supply chain opportunities are challenges that managers in the 21st century must explore and exploit. However, supply chain integration is a means to increased profitability and growth and not an end in itself.

Communication Techniques in Logistics

In an effective supply chain, an increase in the flow of information can result in lower inventories, increased productivity, and improved customer service.

The Communication Technique is the backbone of logistics and effective supply chain management. It is the life-blood of competent logistics performance. Without a sound communication technique it will be almost impossible to seek a competitive advantage through logistics.

Logistics information technique is seen as the main source to improve the productivity and competitive ability, meanwhile reduce the cost. With the development of communication techniques, some items have been already widely used in logistics, such as bar coding, RFID (Radio Frequency Identification), GPS (Global Position Systems), GPRS (General Packet Radio Service), and GIS (Geographic Information Systems).[4]

Bar Code

Information embedded into an identification pattern of parallel bars that can be read by an electronic scanner.

GIS (Geographic Information Systems)

Geographic based software systems used in the transportation industry to optimize route planning, dispatching, facility management, and other supply chain support functions.

GPRS (General Packet Radio Service)

An always-available cellular network technology used to provide Internet and Intranet access.

GPS (Global Positioning Systems)

A group of 24 geo-stationary satellites orbit the earth and provide positioning and telemetry data to ground receivers to pinpoint their geographical location and traveling speed. The location accuracy is in the range 10 – 100 meters for most equipment.

RFID (Radio Frequency Identification)

RFID is an umbrella term for systems that use radio waves to identify items bearing suitably equipped tags. That means the use of radio frequency signals to provide automatic identification of items. One application of radio frequency tracing capabilities is for yard management. Such a system traces vehicles' arrival and departure times, identifies locations and searches yards for lost equipment. Forklifts can be similarly tracked in warehouses. Tags on trucks can be read as far as a quarter of a mile.

The development of the communication technique improves the function of logistics to a great extent. Traditionally, the disadvantages of logistics activities in communication transmission are obvious because of the movement and dispensability during the transportation and material handling. Information and position always change in practice according with the concrete conditions. The communication technique makes it easy to control and manage the logistics activities. The drivers of forklift could get instructions at any times and places in the use of Radio Frequency rather than taking the orders printed before they set out. Consequently, the importance of communication technique cannot be underestimated. It marks the modernization of the logistics.

Notes

1. What began during the last decade of the 20th century and will continue to unfold well into the 21st century is what historians characterized as the dawning of the information or digital age. In the age of electronic commerce, the reality of B2B connectivity has made possible a new order of business relationship called supply chain management.

句意：开始于20世纪最后10年，并将继续繁荣于21世纪的是历史学家们所描绘的信息时代或数字化时代的破晓。在电子商务时代，B2B连通性的现实使得被称作供应链管理的一种商业关系的新秩序成为可能。

2. The supply chain activities encompass all associated with the flow and transformation of goods, the flow of information from the raw materials supplier to the end user, as well as the reverse flows of materials and information in the supply chain.

句意：供应链活动包含所有关于物资的流程和转换环节，从原材料供应者到最终用户的信息流以及供应链中相反的材料和信息流。

3. The integrated supply chain perspective shifts traditional channel arrangements from loosely linked groups of independent businesses that buy and sell inventory to each other toward a managerially coordinated initiative to increase market impact, overall efficiency, continuous improvement, and competitiveness.

句意：集成化供应链从松散的独立商业群体之间互相买卖存货的传统销售渠道模式转变为管理协同的自主性模式，从而增强市场冲击力、综合效率、持续进步能力以及竞争性。

4. With the development of communication techniques, some items have been already widely used in logistics, such as bar coding, RFID (Radio Frequency Identification), GPS (Global Position System), GPRS (General Packet Radio Service), and GIS (Geographic Information System).

句意：随着通信技术的发展，一些技术已经广泛应用于物流中，如条形码、无线射频识别、全球定位系统、通用分组无线业务和地理信息系统等。

Exercises

1. What is supply chain management? And what are the functions of supply chain?

2. Cite some items of logistics information techniques that are widely used.

3. Translate the Following Passages into English:

(1) 然而，太多的顾客服务将不必要地减少公司的利润。对一个公司来说，奉行以顾客需要为基础的顾客服务政策，对整体销售策略的执行始终如一，并促进公司长远利益目标，是必不可少的。

(2) 物流一体化的最新发展是“快速反应”。快速反应将孤立的供应链联系在一起，这样顾客从零售商购买产品后就能自动地将信息反馈给供应链并引发制造商或供应商做出反应。其结果是面向消费者，开发出零售商与供应商之间的伙伴关系，以及供应链成员之间的高度一体化。

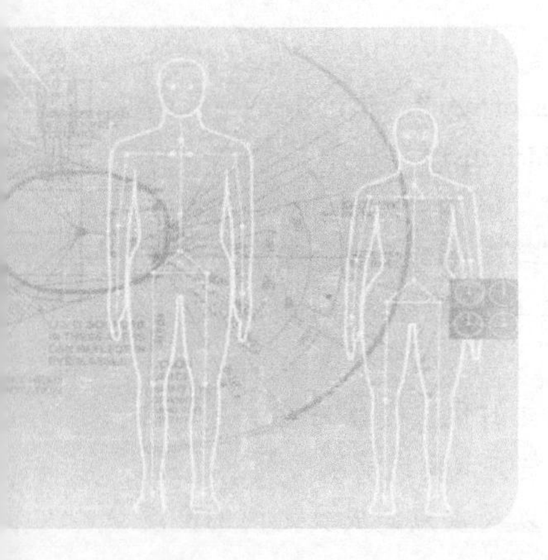

Chapter 7 Ergonomics

Unit 1 Introduction to Ergonomics

Definition of Ergonomics

What is ergonomics (sometimes called human factors)? Most people think it is something to do with seating or with the design of car controls and instruments. It is... but it is much more! Ergonomics is the scientific discipline concerned with the understanding of interactions among humans and other elements of a system, and the profession that applies theory, principles, data and methods to design in order to optimize human being activities and overall system performance. In other words, ergonomics is the application of scientific information concerning humans to the design of objects, systems and environment for human use. Ergonomics is an effective approach which puts human needs and capabilities at the focus of designing technological systems. Until now, ergonomics comes into everything which involves people, such as work systems, sports and leisure, health, safety, and so on.

Aim of Ergonomics

Why do some car seats leave you aching after a long journey? Why do some computer workstations result in eyestrain and muscle fatigue? Such human inconveniences are not inevitable? What is the aim of ergonomics? The aim of ergonomics is to ensure that humans and technology work in complete harmony, with the equipment and tasks aligned to human characteristics.

The ergonomist works in teams which may involve a variety of other professions: design engineers, production engineers, industrial designers, computer specialists, industrial physicians, health and safety practitioners, and specialists in human resources. The overall aim is to ensure that the knowledge of human characteristics is brought to bear on practical problems of people at work and in leisure. In many cases, unsuitable conditions lead often to inefficiency, errors, unacceptable stress, and physical or mental cost.

Ergonomists study human capabilities in relationship to work demands, contribute to the design and evaluation of tasks, jobs, products, environments and systems in order to make them compatible with the needs, abilities and limitations of people. Practicing ergonomists must have a broad understanding of the full scope of the discipline. Ergonomics promotes a complete approach which takes considerations of physical, cognitive, environmental, organizational, social and other relevant factors into account.

Origins of Ergonomics

The term "ergonomics" is derived from the Greek words *ergon* (means *work*) and *nomoi* (means *natural laws*) to express the systems-oriented discipline which now extends across all aspects of human activity.

Ergonomics is a relatively new branch of science which originated in the middle of 20th century, but relies on research carried out in many other older, established scientific areas, such as engineering, physiology and psychology. In World War Ⅱ, when scientists designed advanced new and potentially improved systems without fully considering the people who would be using them, it gradually became clear that systems and products would have to be designed to take many human and environmental factors into account if they are to be used safely and effectively. This awareness of people's requirements resulted in the discipline of ergonomics. The multi-disciplinary nature of ergonomics is immediately obvious.

Components of Ergonomics

Ergonomics deals with the interaction of technological and work situations with human beings. The basic human sciences involved are anthropometry, physiology and psychology. These sciences are applied by the ergonomist towards two main objectives: the most productive use of human capabilities, and the maintenance of human health. In a phrase, the job must 'fit the person' in all aspects, and the work situation should be beyond human capabilities and limitations.

Anthropometry contributes to improve 'physical fit' between people and the things they use, ranging from hand tools to aircraft cockpit design. Achieving good 'physical fit' is no mean feat when one considers the range in human body sizes across the population. Anthropometry provides data on dimensions of the human body, in various postures.

Physiology supports two main technical areas. Work physiology deals with the energy requirements of the body and sets standards for acceptable physical work-rate and workload, and for nutrition requirements. Environmental physiology analyses the impact of physical working conditions—thermal, noise and vibration, and lighting—and sets the optimum requirements for these.

Psychology is concerned with human information processing and decision-making capabilities. In simple terms, this can be seen as aiding the 'cognitive fit' between people and the things they use. Relevant topics are sensory processes, understanding, long- and short-term memory, decision-making and action. The importance of psychological dimensions of ergonomics should not be underestimated in today's ' high-tech' world. The ergonomist advises on the design of interfaces between people and computers, information displays for industrial processes, the planning of training materials, and the design of human tasks and jobs. The concept of information overload is familiar in many current jobs. Increasing automation will frequently increase the mental demands in terms of monitoring, supervision and maintenance.

Research Contents of Ergonomics

Ergonomics includes two aspects: physical ergonomics and cognitive ergonomics.

Physical ergonomics is concerned with human anatomical, anthropometric, physiological and biomechanical characteristics as they relate to physical activity. Relevant topics include working postures, materials handling, repetitive movements, work related musculoskeletal disorders, workplace layout, safety and health.

Cognitive ergonomics is concerned with mental processes, such as understanding, memory, reasoning, and response, as they affect interactions among humans and other elements of a system. Relevant topics include mental workload, decision-making, skilled performance, human-computer interaction, human reliability, work stress and training as these may relate to human-system design.

Underlying all ergonomics work is careful analysis of human activity. The ergonomist must understand all of the demands being made on the person, and the likely effects of any changes to these—the techniques which enable him to do this come under the label of ' job and task analysis'. The second key factor is to understand the users. For example, consumer ergonomics covers applications to the wider contexts of the home and leisure.

Contents of Ergonomic Design

Ergonomic design is a way of considering design options to ensure that people's capabilities and limitations are taken into account. This helps to ensure that the product is fit for the target users. Generally, ergonomic design covers product design, age related design and information design.

Product Design

Even the simplest of products if poorly designed can be a nightmare to use. Human ancestors didn't have this problem. They could simply make things to suit themselves. Nowadays, the designers of products are often far removed from the end users, which makes it vital to adopt an ergonomic, user-centered approach to design, including studying people using equipment, talking to them and asking them to test objects. This is especially important with ' inclusive design' where everyday products are designed with older and disabled users in mind.

Age Related Design

The number of people aged 75 and over is forecast to double over the next 50 years. It is necessary to extend the range of application of equipment, services and systems designed for the general population. Data needs to be available on relevant aspects of the capability of the whole population including older and disabled people. The aspects include the physiological (for instance, range of limb

movement, strength, vision and hearing) and the psychological (for example, cognitive, reaction time, memory). Anthropometric data is also required (size and shape ranges of people). With data such as this available, a knowledge base can be generated for access by responsible designers. Quality of life for older and disabled people may also be enhanced by improvements in the built environment. This includes design of the home, design of public access buildings and public spaces, and design and operation of transport systems. Physical aspects of design that need to be considered include stairs and ramps, comfortable conditions (cold, damp, heat), security and accessibility. Sensory aspects include acoustics, lighting, comfort, communication systems and navigation.

Information Design

Much of today's ergonomics research is channeled towards improving the ways people use information. In fact, everyone has experienced the frustration of using computer software that doesn't work the way they expect it to. For the majority of end users of computer programs, if the system is not working they have no aid but to call for technical help, or find creative ways around system limitations, using the parts that are usable, or increasing stress levels by using a substandard system. Often the problems in systems could have been avoided, if a more complete understanding of the users' tasks and requirements had been present from the start. The development of easily usable human-computer interfaces is a major issue for ergonomists today. Information design is a related area, concerned with the design of signs, symbols and instructions so that their meaning can be quickly and safely understood.

Application of Ergonomics

Ergonomics has a wide application to everyday domestic situations, but there are even more significant implications for efficiency, productivity, safety and health in work settings. For example:

- Designing equipment and systems including computers, so that they are easier to use and less likely to lead to errors in operation;
- Designing tasks and jobs so that they are effective and take account of human needs such as rest breaks and sensible shift patterns, as well as other factors such as essential rewards of work itself;
- Designing equipment and work arrangements to improve working posture and ease the load on the body, thus reducing occasions of Repetitive Strain Injury/Work Related Upper Limb Disorder;
- Designing information, to make the interpretation and use of handbooks, signs, and displays easier and decrease error;
- Designing working environments, including lighting and heating, to suit the needs of the users and the tasks performed.

Exercises

1. What is the definition of ergonomics?
2. What is the aim of ergonomics?
3. What does ergonomics consist of ?
4. What is the main content of Ergonomic Design?

Unit 2 Physiological and Psychological Activities of Human

Physiological Aspects of Human Performance

Ergonomists study human capabilities in relationship to work demands. In recent years, ergonomists have attempted to define postures which minimize unnecessary static work and reduce the forces acting on the body. All of people could significantly reduce the risk of injury if the following ergonomic principles are obeyed:

- All work activities should permit the worker to adopt several different, but equally healthy and safe postures;
- Muscular force has to be exerted by the largest appropriate muscle groups available;
- Work activities should be performed with the joints at about mid-point of their range of movement. This applies particularly to the head, trunk, and upper limbs.

However, in order to put these principles into practice, a person would have to be a skilled observer of his or her own joint and muscle functioning and would have to be able to change his or her posture to a healthier one at will. No one develops this sort of highly refined sensory awareness without special training. Therefore, in order to derive the benefits of ergonomic research, learning to use the data of anthropometry is necessary so that the user can keep a proper posture.

Anthropometry

Anthropometry is a branch of ergonomics that deals with body shape and size. People come in all shapes and sizes so you need to take these physical characteristics into account whenever you design anything that someone will use, from something as simple as a pencil to something as complex as a car. Anthropometry tables give measurements of different body parts for men and women, and split into different nationalities and age groups, from babies to the elderly. So first of all you need to know exactly whom you are designing for. The group of people you are designing for is called the user population.

If you were designing an office chair, you would need to consider dimensions for adults of working age and not those for children or the elderly. If you were designing a product for the home, such as a kettle, your user group would include everyone except young children. You need also to know which parts of the body are relevant to your design. For example, if you were designing a mobile phone, you would need to consider the width and length of the hand, the size of the fingers, as well as handle diameter. You wouldn't be too interested in the height or weight of the user. Besides, you should know whether you are designing for the 'average' or extremes. The variation in the size and shape of people also tells us that if you design to suit yourself, it will only be suitable for people who are the same size and shape as you, and you might 'design out' everyone else!

Percentile

Percentiles are shown in anthropometry tables and they tell you whether the measurement given in the tables relates to the 'average' person, or someone who is above or below average in a certain dimension. If you look at the heights of a group of adults, you'll probably notice that most of them look about the same height. A few may be noticeably taller and a few may be noticeably shorter. This 'same height' will be near the average (called the 'mean' in statistics) and is shown in anthropometry tables as the fiftieth percentile, often written as '50th %ile'. This means that it is the most likely height in a group of people. The graph of the heights (or most other dimensions) of a group of people would look similar to Figure 7-1.

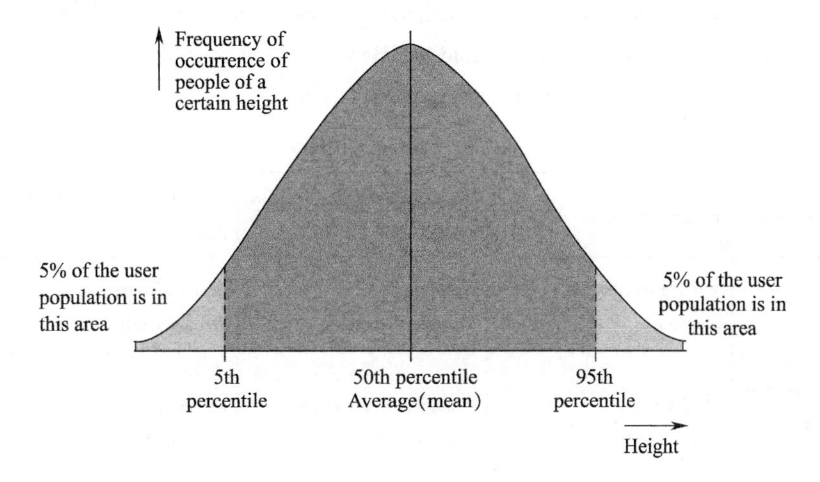

Figure 7-1 The Heights of a Group of People

Notice that the graph is symmetrical so that 50% of people are of average height or taller and 50% of people are of average height or smaller. The graph extends to either end, because fewer people are extremely tall or very short. To the left of the average, there is a point known as the 5th percentile, because 5% of the people (or 1 person in 20) are shorter than this particular height. The same distance to the right is a point known as the 95th percentile, where only 1 person in 20 is taller than this height.

Therefore, designers also need to know whether they are designing for all potential users or just the ones of above or below average dimensions. Now, this depends on exactly what it is that they are designing. For example, if they were designing a doorway using the height, shoulder width, hip width etc., of an average person, then half the people using the doorway would be taller than the average, and half would be wider. Since the tallest people are not necessarily the widest, more than half the users would have to bend down or turn sideways to get through the doorway. Therefore, in this case dimensions of the widest and tallest people are used to design so that everyone could walk through normally.

Application of Anthropometry

Deciding whether to use the 5th, 50th or 95th percentile value depends on what are being designed and *whom* they are designed for. Usually, you will find that if the right percentile is picked, 95% of people will be able to use the design. For instance, if a door height is chosen, the dimension of people's height (often called 'stature' in anthropometry tables) would be chosen and the 95th percentile value be picked—in other words, the design is for the taller people. The average height people are not needed to worry about, or the 5th percentile ones—they would be able to fit through the door anyway.

At the other end of the scale, if an airplane cockpit is designed, it is necessary to make sure everyone could reach a particular control, 5th percentile arm length would be chosen—because the people with the short arms are the ones who are most challenging to design for. If they could reach the control, everyone else (with longer arms) would be able to. Here are some examples of other situations—the design project will normally fit into one of these groups (Table 7-1):

Table 7-1 Design Examples

The aim for the design	Design examples	Examples of measurements to consider	Users that the design should accommodate
Easy reach	Vehicle dashboards Shelving	Arm length Shoulder height	Smallest user: 5th percentile
Adequate clearance to avoid unwanted contact or trapping	Manholes Cinema seats	Shoulder or hip width Thigh length	Largest user: 95th percentile
A good match between the user and the product	Seats Cycle helmets Pushchair	Knee-floor height Head circumference Weight	Maximum range: 5th to 95th percentile
A comfortable and safe posture	Lawnmowers Monitor positions	Elbow height Sitting eye height	Maximum range: 5th to 95th percentile
Easy operation	Screw bottle tops Door handles Light switches	Grip strength Hand width Height	Smallest or weakest user: 5th percentile
To ensure that an item can't be reached or operated	Machine guarding mesh Distance of railings from hazard	Finger width Arm length	Smallest user: 5th percentile Largest user: 95th percentile

Sometimes the design can't accommodate all the users because there are conflicting solutions to the design. In this case, it is necessary to make a judgment about what is the most important feature. The

design must never compromise safety, and if there is a real risk of injury, more extreme percentiles (1%ile or 99%ile or more) will be used to make sure that everyone is protected (not just 95% of people).

Impacts of Working Environment

If tools for changing car wheels are designed, it's more than likely that they would have to be used in cold and wet weather. People need to grip harder if their hands are wet and cold, and they need to exert more force to carry out tasks than they would if they were warm and dry. That is, the impacts of working environment on physiological status must be taken into account.

People's eyesight and hearing abilities are also needed to consider. Can they read the small labels on the remote control that you've designed? Is there enough light to read them by? Can they hear an alarm bell above the general noise in the room? The research work in the area of ergonomics consists of impacts of environment-lighting, noise, vibration and so on. The interactions of these stimulates are factored in the developmental, operational test & evaluations and product design.

Vision is usually the primary channel for information, yet systems are often so poorly designed that the user is unable to see the work area clearly. Many workers using computers cannot see their screens because of glare or reflections. Others, doing precise assembly tasks, have insufficient lighting and suffer eyestrain and reduced output as a result.

Sound can be a useful way to provide information, especially for warning signals. However, care must be taken not to overload this sensory channel. A recent airliner had 16 different audio warnings, far too many for a pilot to deal with in an emergency situation. A more sensible approach was to have just a few audio signals to alert the pilot to get information guidance from a visual display.

Definition and Goal of Engineering Psychology

Engineering Psychology is defined as the application of psychological principles, knowledge, and research to improve the ability of humans to operate more effectively in a technological society. General research focuses on people's interaction with or involvement with communication, decision making, and computer information systems, work places, energy and transportation systems, medical and health care settings, consumer product design, living environments, etc.

The goal of Engineering Psychology is safer, more effective, and more reliable systems through improved understanding of the user's requirements and performance capabilities. This aspect of psychology is often referred to as "ergonomic psychology", and the goal of ergonomics is defined as making the human interaction with systems one that reduces error, increases productivity, and enhances safety and comfort.

Effects of Working Environment on Safety

Safety and accident prevention are important areas of ergonomics, environmental psychology and ergonomics engineering. What effect does personality have on accident rates? Can eyestrain and mental stress injuries be reduced by improving the design of computer displays and input devices? Are some types of office furniture benefit to increased productivity? What design features of traffic lights have optimal visibility across all climatic conditions? How can people be reminded to use seat belts? How can roads, sidewalks, and crosswalks be designed to increase walker safety? What is the effect of mobile telephones on automobile-accident rates? What are good criteria for choosing candidates for flight training? Through how many working hours can an air traffic controller remain on the alert?

Our surroundings affect our sense, our work, our learning, our recovery from illness, and our mental health. Psychologists study the effect on individuals of crowding versus open spaces, of rapid social and technological change, of changing jobs and relocating families, of living in high-rises and underground and in submarines, and of colors and windows. The results of this research are applied in urban and suburban planning, in the design of homes, hospitals, and schools, and in organizational and governmental policy making.

The general surroundings of your office could be having negative effects on your mental state and work productivity. The condition of your office furnishings and the equipment you have to use can play a very important role in how much you enjoy your day and how effective you are as an employee. Psychologically your office has a big effect on you.

Environmental Factors

It is essential that as an employee you receive a suitable amount of natural light when sitting at your desk and when anywhere in your office. You must also be able to control the amount of natural light in the office with the use of blinds or shades. Apart from natural light, electric light can have a major effect. Strong overhead lighting that can cause headaches, eyestrain and fatigue can be reduced by simply adding filters or introducing lower indirect lighting. Eyestrain and headaches can appear as a result of the glare you get on your monitor screen. Implementing such lighting and shading will enable you to reduce the uneasiness. A lighting scheme suitably designed for a working environment can enhance the safety and operational effectiveness of our servicemen and women. Lighting specialists support infrastructure and platform projects by evaluating and designing lighting schemes for optimal surrounding and task lighting. Recommendations are also made on color schemes in man-made or artificial environments based on man's psychological reactions to color and light.

The noise you are subjected to and the amount of privacy you have can play a very important role in how much you enjoy your day and how effectively you work. Ideally, if your office is open plan there needs to be somewhere you can go to have privacy, be it a separate office or meeting room. Noise can become very stressful as it causes distraction, if it's possible, have acoustical panels fitted to absorb the noise. Noise and vibration are environmental hazards, which can have a negative effect on our servicemen and women's health, combat performance and readiness. Ergonomists support various military platform projects by evaluating and recommending measures to eliminate or reduce the effects of noise and vibration and also research their interactions in enclosed environments.

Mental Health

In some industries the impact of human errors can be miserable. These include the nuclear and chemical industries, rail and sea transport and aviation, including air traffic control. When disasters occur, the blame is often laid with the operators, pilots or drivers concerned and labeled 'human error'. Often though, the errors are caused by poor equipment and system design. Ergonomists working in these areas pay particular attention to the mental demands on the operators, designing tasks and equipment to minimize the chances of misreading information or operating the wrong controls.

High levels of information load are sometimes associated with stress, which in the long term brings up a variety of unhealthy behaviors and states. On the other hand, people have a need for information and are generally well-equipped to deal with large amounts of it. Indeed they can suffer from understimulation, and tire without information. Although the available information in any domain is potentially infinite depending on what details are attended to, perception is selective and in this sense we cannot get overloaded in the same sense as machines can.

Human/Machine System Interface

The machine's main interface, in terms of dialogue and communication between it and the operative, is in the form of two control panels located at strategic points. Some consideration had gone into the screen design and dialogue interaction, with an obvious attention to established usability design criteria. It is understood that numerous simulations were run to fine-tune the final design and achieve the required level of operative ability to respond to, diagnose and correct malfunctions, through to quality non-accordance.

Although limited in extent, those points where the operative is required to physically interact with the machine, had received considerable attention in the context of primary anthropometrics—the body's and limb's principal dimensions. The feed-on/off or loading/unloading points to the machine were designed to minimize awkward or stressful manual handling postures, with optimal heights being chosen for the transmitting systems. The control panels are located at optimal height and orientation with adjustable tilt and rotate functions. Some attention had also gone into the display screen's design, addressing issues such as minimization of screen glare and luminous contrasts.

Notes

The noise you are subjected to and the amount of privacy you have can play a very important role in how much you enjoy your day and how effectively you work.

句意：噪声的影响和私有空间的大小是影响你心情好坏和工作效率高低的重要原因。

be subjected to 意为“受制于……的，受……影响的，以……为条件的”。

Exercises

1. What is the definition of Anthropometry?

2. Generalize the applications of Anthropometry.

3. What are the anthropometric data and ergonomics in relation to cab design?

4. What is the definition of Engineering Psychology?

5. How can the musical instrument the flute be at a suitable size for both a 7 years old and a 17 years old without any extra joints?

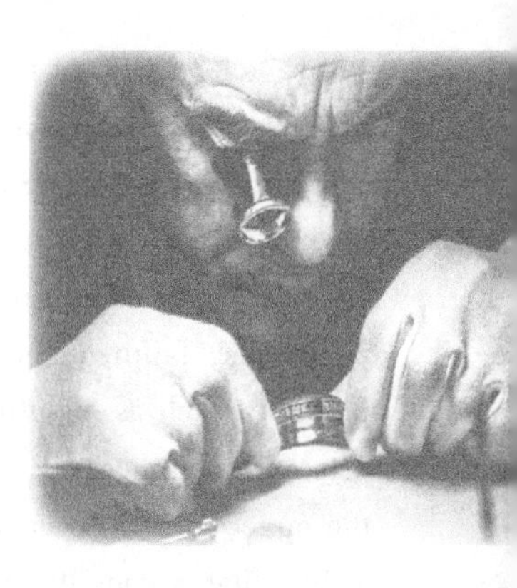

Chapter 8
Quality Management

Unit 1 Quality Standards and Quality Control

Definitions of Quality

Quality is an abstract concept in business. Many people think of quality as some level of superiority or inborn excellence; others view it as a lack of manufacturing defects. The formal definition of quality, standardized by the American National Standards Institute (ANSI)[1] and the American Society for Quality Control (ASQC)[2] in 1978 is "the totality of features and characteristics of a product or service that bears on its ability to satisfy given needs." This definition implies that people must be able to identify the features and characteristics of products and services that determine customer satisfaction and form the basis for measurement and control. The "ability to satisfy given needs" reflects the value of the product or service to the customer, including the economic value, safety, reliability, and maintainability. Complete quality includes two aspects: Fitness for use and conformance to specifications.

Fitness for Use

Although the ANSI/ASQC definition of quality is operationally useful, it does not completely describe the various viewpoints of quality that are commonly used: since customer needs must be the driving force behind quality products and services, a popular definition of quality is fitness for use. This is encompassed in the ANSI/ASQC definition as "the ability to satisfy given needs." This definition means that a quality product or service must meet customer requirements and expectations. The fitness for use definition based on customer satisfaction has become the principal definition of quality from a managerial perspective. By the end of the 1980s, a different definition of quality had emerged: quality is meeting or exceeding customer expectations.

Conformance to Specifications

A second approach to defining quality, from the perspective of manufacturing or service delivery, is accorded with specifications. Specifications are targets and tolerances determined by designers of products and services. Targets are the ideal values for which production is expected to strive; tolerances are acceptable deviations from these ideal values, recognizing that it is impossible to meet the targets all the time. The traditional manufacturing view of quality as conformance to specifications has come under much careful check in recent years because of the work of Japanese engineer Genichi Taguchi[3]. Taguchi defines quality as the avoidance of "the loss a product causes to society after being shipped." This includes losses due to a product's failure to meet customer expectations, failure to meet performance characteristics, and harmful side effects caused by the product, such as pollution or noise. Taguchi measures loss in monetary units and relates it to targets and tolerances. He has shown that the loss increases more rapidly the further one moves from the target value in a critical specification. For example, just think of what would happen if many airline flights consistently varied from scheduled arrival times. The more flights are delayed, the more passengers will miss connecting flights, causing substantial losses to both passengers and the airlines. Variation in the production of products and services cannot be totally eliminated; the variation around target values can be minimized. This minimizes the economic loss and benefits both the producer and the consumer.

Fitness for use (quality of design) and conformance to specifications (quality of conformance) provide the fundamental basis for managing operations to produce quality products. A "customer-driven" quality focus involves every one in an organization. Customer requirements must be determined and understood. They must be translated into detailed product and process specifications. Manufacturing and service delivery must meet these specifications during production to ensure that what the customer gets is what the customer wants (or more). Quality is everyone's responsibility.

Quality in Manufacturing

In manufacturing, quality is an important component of all functions. For example, effective market research is necessary to determine customer needs and identify functional requirements for product designers. Product designers must take care to neither overengineer (resulting in inefficient use of a firm's resources) nor underengineer products (resulting in poor quality). Purchasing must ensure that suppliers meet quality requirements. Production planning and scheduling should not put unsuitable pressure on manufacturing that will degrade quality. Tool engineering and maintenance are responsible for ensuring that tools, gauges, and equipment are properly maintained. Industrial engineering must select the appropriate technology that is capable of meeting design requirements and developing appropriate work methods. Packaging, shipping, and warehousing have the responsibility of ensuring the condition, availability, and timely delivery of products. Accessory functions such as finance, human resources, and legal services support the quality effort by providing realistic budgets, a well-trained and motivated workforce, and reviews of warranty, safety, and liability issues.

Quality in Service

In services, the importance of quality cannot be underestimated. Service is a social act which takes place in direct contact between the customer and representatives of the service company. In serv-

ices, the distinguishing features that determine quality differ from manufacturing. The most important dimensions of service quality include:

- Time: How long must a customer wait?
- Timelines: Will a package be delivered by 9:30 the next morning?
- Completeness: Are all items in the order included?
- Courtesy: Do front-line employees greet each customer cheerfully?
- Consistency: Are services delivered in the same fashion for every customer?
- Accessibility and convenience: Is the service easy to obtain?
- Accuracy: Is the service performed right the first time?
- Responsiveness: Can service personnel react quickly and resolve unexpected problems?

Many service organizations such as airlines, banks, and hotels have well-developed quality assurance systems. Most of them, however, are generally based on manufacturing similarities and tend to be more product-oriented than service-oriented. For example, a typical hotel's quality assurance system is focused on technical specifications such as properly made-up rooms. However, service organizations have special requirements that manufacturing systems cannot fulfill. Service organizations must look beyond product orientation and pay significant attention to customer transactions and employee behavior.

ISO 9000 Series Standards

The ISO 9000 series of quality management standards was developed by the International Organization for Standardization (ISO)[4] to set international requirements for quality management systems in 1979. The series is designed to be applicable to any manufacturing or service process. It is revised periodically and controlled by Technical Committee (TC) 176, made up of international members from many industries and backgrounds. The original standards were published in 1987, first revised in 1994 and the current versions were issued in 2008.

The ISO 9000 series of standards sets out to create a framework of the fundamental elements that would form the basis for a series of internationally recognized quality management standards. It represents the essential requirements that every enterprise needs to address to ensure the consistent production and timely delivery of its goods and services to the marketplace. These requirements make up the standards that contain the quality management system, and the general nature allows for their application in any type of organization.

The ISO 9000 series is able to provide these quality management benefits to any organization of any size, public or private, without dictating how the organization is to be run. The system standards describe what requirements need to be met, not how they are to be met. This allows for diverse organizations to apply the same standards in a manner that reflects the reality of their business structure. In essence, allowing each organization to meet the system requirements by implementing the standards in a manner that suits its own unique needs. Increasingly, certification to an internationally recognized quality management standard like one from the ISO 9000 series is becoming an important part of distinguishing an organization from its competition.

Rules of Certification to QMS

Certification of a Quality Management System (QMS) to the generic ISO 9001: 2008 standard can improve the ability to meet strategic objectives. The road to certification will help to prepare for an independent audit. Certification, or registration, to a standard is the outcome of a successful assessment by an independent third party. Experience has shown that some simple tips, often seems to be trivial, have proven invaluable to companies seeking certification. The following rules should be obeyed:

- Make sure you begin the process with the right attitude;
- Have a complete understanding of the concept set forth in the standard, and use the standard as a guide template to define your management system;
- Know what application and implications of the standard will mean to your company;
- Use the standard as a tool for improvement;
- Have an understanding of the risks and processes that affect your organization's ability to realize its business strategy;
- Select your partner (certification body/registrar) carefully.

Steps of Certification to QMS

The following 10 general steps is the road to certification:

Obtain a Standard

Obtain and read a copy of the standard to familiarize yourself with the requirements. Then you should decide if certification/registration to this standard makes good sense for your organization.

Review Literature and Software

There is a large amount of published information available that is designed to assist you in understanding and implementing a standard.

Assemble a Team and Define Your Strategy

The adoption of a management system needs to be the strategic decision of the whole organization. It is vital that your senior management is involved in the creation process. They decide the business strategy that an efficient management system should support. In addition, you need a dedicated team to develop and implement your management system.

Determine Training Needs

Your team members responsible for implementing and maintaining the management system(s) will need to know the full details of the applicable standard(s). There is a wide range of courses, workshops, and seminars available designed to meet these needs.

Review Consultant Options

Independent consultants will be able to advise you of a workable, realistic, and cost effective strategy plan for implementation.

Develop a Management Systems Manual

Your management systems manual should describe the policies and operations of your company.

Through the manual, you will provide an accurate description of the organization and the best practice adopted to consistently satisfy customer expectations.

Develop Procedures

Procedures describe the processes of your organization, and the best practice to achieve success in those processes. These procedures should answer the following questions about each process: why, who, when, where, what and how.

Implement Your Management System

Communication and training are key to a successful implementation. During the implementation phase, your organization will be working according to the procedures that were developed to document and demonstrate the effectiveness of the management system.

Consider a Pre-assessment

You can choose to have a preliminary evaluation of the implementation of your management system by a certification body/registrar. The purpose of this is to identify areas of non-accordance and allow you to correct these areas before you begin the qualified certification process. Receive a non-conformance means that a particular area of your management systems is not suitable to the requirements of the standard.

Select a Certification Body/Registrar

The business relationship with the certification body/registrar will be in place for many years, as the certification has to be maintained. In this age of corporate check in detail, it is forcible to choose a certification body/registrar with a reputation beyond reproach. When the management system is implemented, prepared for certification, and the certification body/registrar is chosen, the accredited certification can be begun.

Statistic Quality Control

Quality control involves making a series of planned measurements in order to determine if quality standards are being met. If not, then corrective action and future preventive action must be taken to achieve and maintain accordance. Quality control activity may be required at times or in places where supervision and control personnel are not present. Work must often be performed at the convenience of the customer. Hence, more training of employees and self-management are necessary.

Definition of Six Sigma Methodology

Six Sigma (Sigma is the Greek letter used to represent standard deviation[5] in statistics) stands for Six Standard Deviations from mean. Six Sigma methodology provides the techniques and tools to improve the capability and reduce the defects in any process. It was started in Motorola, in its manufacturing division, where millions of parts are made using the same process repeatedly. Finally Six Sigma evolved and applied to other non-manufacturing processes. Today Six Sigma can be applied to many fields such as services, medical and insurance procedures and call centers.

Six Sigma methodology improves any existing business process by constantly reviewing and re-tuning the process. To achieve this, Six Sigma uses a methodology known as DMAIC (Define opportunities, Measure performance, Analyze opportunity, Improve performance, Control performance). Six Sigma methodology can also be used to create a brand new business process from ground up using DFSS (Design For Six Sigma) principles. It allows for only 3.4 defects per million opportunities for each product or service transaction. Six Sigma relies heavily on statistical techniques to reduce defects and measure quality. Six Sigma experts evaluate a business process and determine ways to improve upon the existing process. Six Sigma experts can also design a brand new business process using DFSS principles. Typically it's easier to define a new process with DFSS principles than refining an existing process to reduce the defects.

Six Sigma incorporates the basic principles and techniques used in business, statistics, and engineering. These three areas form the core elements of Six Sigma. Six Sigma improves the process performance, decreases variation and maintains consistent quality of the process output. This leads to defect reduction and improvement in profits, product quality and customer satisfaction.

Key Elements of Six Sigma Methodology

Customer requirements, design quality, metrics and measures, employee involvement and continuous improvement are main elements of Six Sigma Process Improvement. The key elements of Six Sigma are: Customer Satisfaction, Defining Processes, Defining Metrics and Measures for Processes, Using and Understanding Data and Systems, Setting Goals for Improvement, and Team Building and Involving Employees.

Involving all employees is very important to Six Sigma. The company must involve all employees. Company must provide opportunities and motivations for employees to focus their talents and ability to satisfy customers.

Six Sigma in Business

Even though Six Sigma was initially implemented at Motorola to improve the manufacturing process, all types of businesses can profit from implementing Six Sigma. Businesses in various industry segments such as services industry (Example: Call Centers, Insurance, Financial/Investment Services), e-commerce industry, education can definitely use Six Sigma principles to achieve higher quality. Many big businesses such as General Electric Co. GE and Motorola have successfully implemented Six Sigma but the adaptation by smaller businesses has been very slow. GE is a pioneer in using Six Sigma. This charpter on Six Sigma GE Experiences explains how various GE divisions adopted and benefited from Six Sigma. Here are some of the reasons to consider:

Bigger companies have resources internally who are trained in Six Sigma and also have 'Train the Trainer' programs using which they cultivate many Six Sigma instructors. Also many bigger companies encourage the employees to learn Six Sigma process by providing Green Belts/Black Belts[6] as instructors. Big companies make Six Sigma as part of the Goals for employees and provide incentives for employees who undergo training and advisor colleagues. Many assume that Six Sigma works for bigger companies only as they produce in volumes and have thousands of employees. This notion is not true and Six Sigma can be effectively applied for small businesses and even companies with fewer than 10 employees.

Six Sigma in Engineering

A Six Sigma Engineer develops efficient and cost effective processes to improve the quality and reduce the number of defects per million parts in a manufacturing/production environment. Six Sigma Engineers determine and fine tune manufacturing process. Once a process is improved, they go back and re-tune the process and reduce the defects. This cycle is continued till they reach 3. 4 or less defects per million parts. Six Sigma is all about knowledge sharing. If a company has more than one manufacturing unit/plant, it's more than likely that one of the plants produces better quality than others. The Six Sigma team should visit this higher quality plant and learn why its perform better than others and implement the techniques learned across all other units. Research/Design (R&D) department within a company can use the above techniques to learn from another R&D department in the same company or affiliate companies and implement those techniques.

Quality Control Tools

Production environments that utilize modern quality control methods are dependant upon statistical literacy. The tools used therein are called the seven quality control tools. These include:

Checksheet

The function of a checksheet (Figure 8-1) is to present information in an efficient, graphical format. This may be accomplished with a simple listing of items. However, the utility of the checksheet may be significantly enhanced, in some instances, by incorporating a description of the system under analysis into the form.

Sample Checksheets

Defect Type		Totals
1. Assembly	II	2
2. Print Quality	IIIIIIIIIIIII	13
3. Print Detail	IIII	4
4. Edge Flaw	IIIIIIIIIIIIIIIIIIIIII	22
5. Cosmetic	IIIII	5

Customer Complaints		Totals
1. Missing Ring	II	2
2. Print Quality	IIIIIIIIIIIIIIIIIIIIIII	23
3. Misplace Print	IIII	4
4. Rough Edge	III	3
5. Type Error	IIIIII	6
6. Excess Flash	IIIIIIIIIIIII	13
7. Late Shipment	IIIIII	6
8. Bad Count	IIII	4

Figure 8-1 Checksheets

Pareto Chart

Pareto charts[7] (Figure 8-2) are extremely useful because they can be used to identify those factors that have the greatest cumulative effect on the system, and thus screen out the less significant factors in an analysis. Ideally, this allows the user to focus attention on a few important factors in a process. They are created by plotting the accumulated frequencies of the relative frequency data (event count data), in descending order. When this is done, the most essential factors for the analysis are graphically apparent, and in an orderly format.

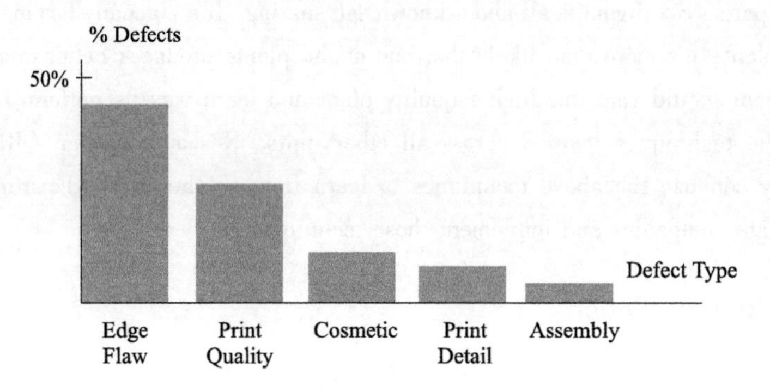

Figure 8-2 Pareto Chart

Flowchart

Flowcharts (Figure 8-3) are diagrammatic representations of a process. By breaking the process down into its necessary steps, flowcharts can be useful in identifying where errors are likely to be found in the system.

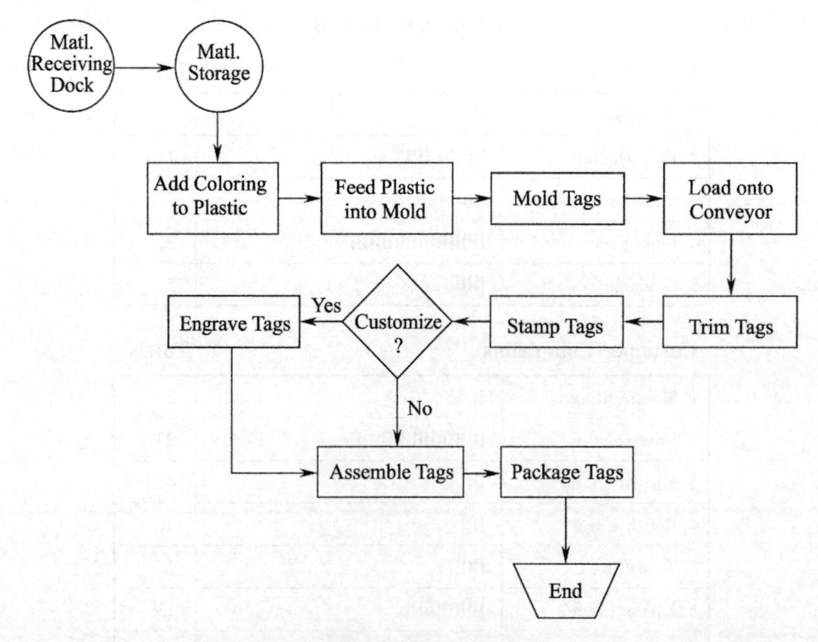

Figure 8-3 Plastic Tag Production Flowchart

Cause and Effect Diagram

Cause and Effect Diagram[8] (Figure 8-4), also called a fish bone diagram, is used to associate multiple possible causes with a single effect. Thus, given a particular effect, the diagram is constructed to identify and organize possible causes for it.

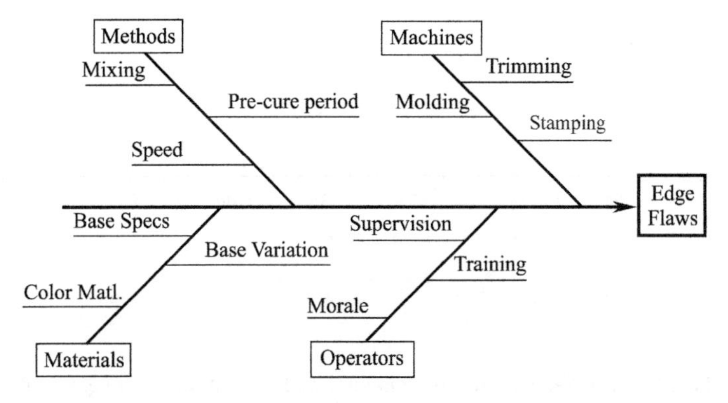

Figure 8-4 Cause and Effect Diagram for Edge Flaws

The primary branch represents the effect (the quality characteristic that is intended to be improved and controlled) and is typically labeled on the right side of the diagram. Each major branch of the diagram corresponds to a major cause (or class of causes) that directly relates to the effect. Minor branches correspond to more detailed causal factors. This type of diagram is useful in any analysis, as it illustrates the relationship between cause and effect in a rational manner.

Histogram

Histograms (Figure 8-5) provide a simple, graphical view of accumulated data, including its dispersion and central tendency. In addition to the ease with which they can be constructed, histograms provide the easiest way to evaluate the distribution of data.

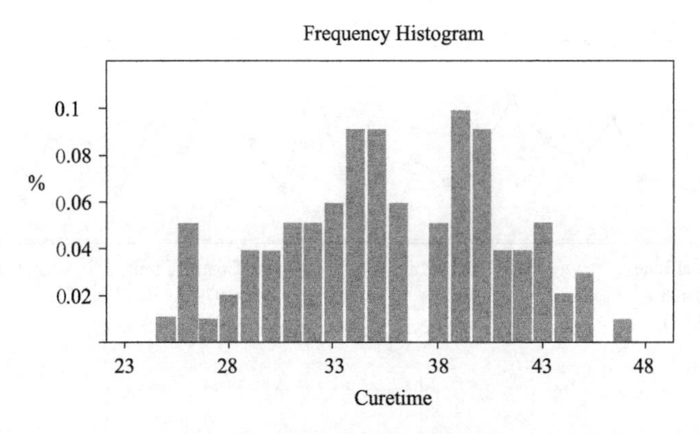

Figure 8-5 Histograms

Scatter Diagram

Scatter diagrams (Figure 8-6) are graphical tools that attempt to describe the influence that one variable has on another. A common diagram of this type usually displays points representing the observed value of one variable corresponding to the value of another variable.

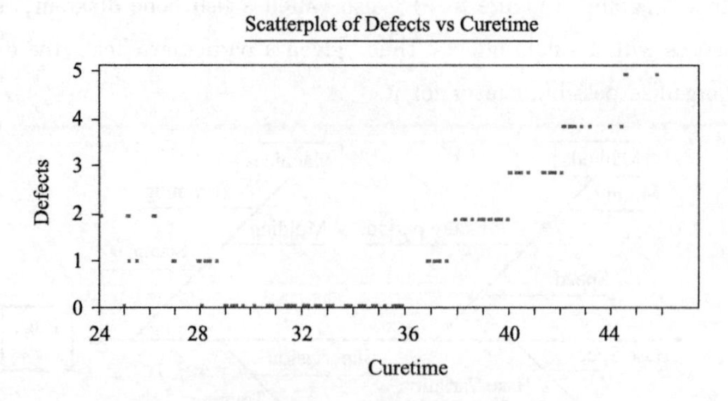

Figure 8-6 Scatter Diagram

Control Chart

The control chart (Figure 8-7) is the fundamental tool of statistical process control, as it indicates the range of variability that is built into a system. Thus, it helps determine whether or not a process is operating consistently or if a special cause has occurred to change the process mean or variations. The bounds of the control chart are marked by upper and lower control limits that are calculated by applying statistical formulas to data from the process. Data points that fall outside these bounds represent variations due to special causes, which can typically be found and eliminated. On the other hand, improvements in common cause variation require fundamental changes in the process.

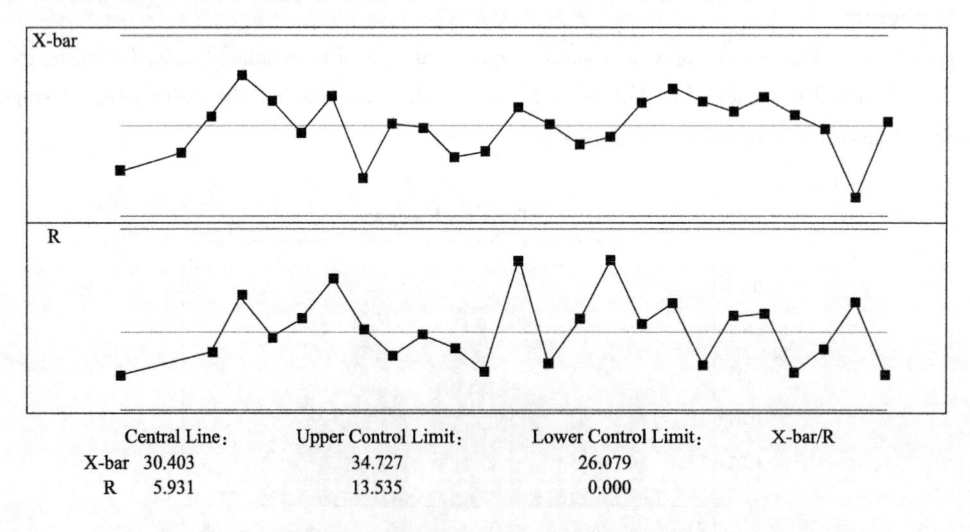

Figure 8-7 Control Chart for Statistical Process

The tools listed above are ideally utilized in a particular methodology, which typically involves either reducing the process variability or identifying specific problems in the process. However, other methodologies may need to be developed to allow for sufficient customization to a certain specific process. In any case, the tools should be utilized to ensure that all attempts at process improvement include: discovery, analysis, improvement, monitoring, implementation and verification. Furthermore,

it is important to note that the mere use of the quality control tools does not necessarily constitute a quality program. Thus, to achieve lasting improvements in quality, it is essential to establish a system that will continuously promote quality in all aspects of its operation.

Notes

1. American National Standards Institute (ANSI) 美国国家标准化组织
2. American Society for Quality Control (ASQC) 美国质量控制学会
3. Genichi Taguchi 田口玄一，日本著名的质量专家，创立了“质量工程学”
4. International Organization for Standardization (ISO) 国际标准化组织
5. standard deviation 标准偏差，标准方差
6. Green Belts/Black Belts 绿带/黑带，表示质量管理人员级别的标志
7. Pareto charts 帕累托图，直条构成的线图，又叫排列图
8. Cause and Effect Diagram 因果图

Exercises

1. What is quality? What is the difference between quality in manufacturing system and in service system?
2. What is the usage of different quality control tools?
3. What does Green Belts stand for? What is the difference between the Green Belts and Black Belts?
4. What kind of information can we get from a histogram?
5. How to get a control chart?

Unit 2 Quality Management and Quality Cost

Definition of TQM

Quality management involves the planning, organization, direction, and control of all quality assurance activities. While many manufacturing firms have quality control departments to provide technical support, successful businesses have found that quality must be integrated throughout the firm.

Total Quality Management (TQM)[1] is a management philosophy that seeks to integrate all organizational functions (marketing, finance, design, engineering, and production, customer service, etc.) to focus on meeting customer needs and organizational objectives. TQM is a method by which management and employees can become involved in the continuous improvement of the production of goods and services. It is a combination of quality and management tools aimed at increasing business and reducing losses due to wasteful practices. Total Quality is a description of the culture, attitude and organization of a company that strives to provide customers with products and services that satisfy their needs. The culture requires quality in all aspects of the company's operations, with processes being done right the first time and defects and waste eradicated from operations. TQM is originated in the 1950's and has steadily become more popular since the early 1980's. Many famous companies, such as Ford Motor Company, Phillips Semiconductor, SGL Carbon, Motorola and Toyota Motor Company have implemented TQM.

TQM views an organization as a collection of processes. It maintains that organizations must strive to continuously improve these processes by incorporating the knowledge and experiences of workers. The simple objective of TQM is "Do the right things, right the first time, every time". TQM is infinitely variable and adaptable. Although originally applied to manufacturing operations, and for a number of years only used in that area, TQM is now becoming recognized as a generic management tool, just as applicable in service and public sector organizations. There are a number of evolutionary strands, with different sectors creating their own versions from the common ancestor.

This shows that TQM must be practiced in all activities, by all personnel, in Manufacturing, Marketing, Engineering, Research & Development, Sales, Purchasing, Human Resource, etc.

Principles of TQM

The key principles of TQM are as following:

Management commitment. Plan (drive, direct), Do (deploy, support, participate), Check (review), Act (recognize, communicate, revise).

Employee empowerment. Training, suggestion scheme, measurement and recognition, excellence teams.

Fact based decision making. SPC (Statistical Process Control), The 7 statistical tools, TOPS (Team Oriented Problem Solving).

Continuous improvement. Systematic measurement and focus on excellence teams, cross-functional process management, attain, maintain, improve standards.

Customer focus. Supplier partnership, service relationship with internal customers, never compromise quality, customer driven standards.

The Concept of Continuous Improvement by TQM

TQM is mainly concerned with continuous improvement in all work, from high level strategic planning and decision-making, to detailed execution of work elements on the shop floor. It comes from the belief that mistakes can be avoided and defects can be prevented. It leads to continuously improving results, in all aspects of work, as a result of continuously improving capabilities, people, processes, technology and machine capabilities.

Continuous improvement must deal not only with improving results, but more importantly with improving capabilities to produce better results in the future. The five major areas of focus for capability improvement are demand generation, supply generation, technology, operations and people capability. A central principle of TQM is that mistakes may be made by people, but most of them are caused, or at least permitted, by faulty systems and processes. This means that the root cause of such mistakes

 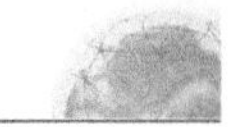

can be identified and eliminated, and repetition can be prevented by changing the process. There are three major mechanisms of prevention:

- Preventing mistakes (defects) from occurring (Mistake-proofing);
- Where mistakes can't be absolutely prevented, detecting them early to prevent them being passed down the value added chain (Inspection at source or by the next operation);
- Where mistakes recur, stopping production until the process can be corrected, to prevent the production of more defects (Stop in time).

Two Implementation Approaches of TQM

A preliminary step in TQM implementation is to assess the organization's current reality. Relevant preconditions have to do with the organization's history, its current needs, precipitating events leading to TQM, and the existing employee quality of working life. If the current reality does not include important preconditions, TQM implementation should be delayed until the organization is in a state in which TQM is likely to succeed.

Traditional Management Approach

This is the most common. This approach represents the 80% failure of TQM's. In this approach TQM never becomes an accepted reality by either organizational or human resource management. It is usually seen as competition, or "something to be tolerated." The TQM system consumes valuable resources needed by the other systems and rejection begins to occur.

Integrated Management Approach

This is the least common. A TQM is blended and balanced with existing cultural backgrounds in both organizational and human resource management systems. This represents the 20% success rate of TQM's. Whether both organizational management and human resource management systems take on a "quality management commitment" or "join a quality management team" is not important. The principles of quality management are attended to as an important third system that blends, integrates, aligns and maximizes the other two systems to beat competition in world class quality performance. This approach can often be divided into two sub-choices, depending upon managerial resources, readiness, acceptance, and abilities.

Definition and Contents of Quality Costs

What is the Quality Costs? Does it raise the price of goods and services? Are huge savings possible by implementing continual improvement efforts? These questions are not easy ones, but quality is measurable, as are its costs. Quality costs are the costs associated with preventing, finding, and correcting defective work. These costs are huge, running at 20%-40% of sales. Many of these costs can be significantly reduced or completely avoided. One of the key functions of a Quality Engineer is the reduction of the total cost of quality associated with a product.

In the 1950s, A. V. Feigenbaum, then Vice President at General Electric (GE), developed and implemented the "Quality Cost" concept throughout GE. He divided Quality Cost into Costs of Control (including Prevention Costs and Appraisal Costs[2]) and Costs of Failure of Control (including Internal Failure Costs and External Failure Costs).

Prevention Costs are costs of activities that are specifically designed to prevent poor quality. Examples of "poor quality" include coding errors, design errors, mistakes in the user manuals, as well as badly documented complex code. Note that most of the prevention costs don't fit within the Testing Group's budget. This money is spent by the programming, design, and marketing staffs. Prevention costs include: staff training, requirements analysis, early prototyping, fault-tolerant design, defensive programming, usability analysis, clear specification, accurate internal documentation, and evaluation of the reliability of development tools or of other potential components of the product.

Appraisal Costs are costs of activities designed to find quality problems, such as code inspections and any type of testing. Design reviews are part prevention and part appraisal. To the degree that you're looking for errors in the proposed design itself when you do the review, you're doing an appraisal. To the degree that you are looking for ways to strengthen the design, you are doing prevention.

Appraisal costs include: design review, code inspection, glass box testing, black box testing, training testers, beta testing, test automation, usability testing, and pre-release out-of-box testing by customer service staff.

Internal Failure Costs are failure costs that arise before your company supplies its product to the customer. Along with costs of finding and fixing bugs are many internal failure costs borne by groups outside of Product Development. If a bug blocks someone in your company from doing her job, the costs of the wasted time, the missed milestones, and the overtime to get back onto schedule are all internal failure costs. Internal failure costs include: bug fixes, regression testing, wasted in-house user time, wasted tester time, wasted writer time, wasted marketer time, wasted advertisements, direct cost of late shipment, and opportunity cost of late shipment.

External Failure Costs are failure costs that arise after your company supplies the product to the customer, such as customer service costs, or the cost of patching a released product and distributing the patch. External failure costs are huge. It is much cheaper to fix problems before shipping the defective product to customers. External failure costs include: technical support calls, preparation of support answer books, investigation of customer complaints, refunds and recalls, shipping of updated product, added expense of supporting multiple versions of the product in the field, lost sales, lost customer goodwill, discounts to resellers to encourage them to keep selling the product, warranty costs, liability costs, government investigations, penalties, and all other costs imposed by law.

Development of Quality Costs

In the 1960s, Dr. James Harrington was assigned to implement Dr. Feigenbaum's concept of Quality Cost at IBM. He found the concept lacking because it did not focus on the support functions' cost of quality or the external customers' quality costs. As a result, IBM expanded the concept to fulfill its own needs and used the name "Poor-Quality Cost (PQC)". This name was selected because IBM felt that Quality Cost was an inappropriate title: Good quality does not cost any additional money. It is poor quality that generates the additional costs for the company. If we had perfect quality, we would

not have a need for preventive cost or appraisal cost, and there would be no internal error cost or external error cost. Unfortunately, we do not live in a perfect world.

During the 1970s, a number of technical reports and articles documenting IBM's approach to Poor-Quality Cost were published. IBM's Poor-Quality Cost system included direct poor-quality cost (Controllable poor-quality cost—prevention cost and appraisal cost, Resultant poor-quality cost—internal error cost and external error cost, Equipment poor-quality cost) and indirect poor-quality cost (Customer-incurred cost, Customer-dissatisfaction cost, Loss-of-reputation cost).

In the early 1980s, Philip Crosby left International Telephone and Telegram Corporation (IT&T) where he had earlier also implemented Feigenbaum's concept of Quality Cost and incorporated the concept into his own consulting practice, he wrote that the quality cost is "the expense of nonconformance—the cost of doing things wrong" in "Quality is Free". Crosby's approach closely followed Feigenbaum's teachings, although he changed the names of some of Feigenbaum's terms without altering their meaning.

In 1994, after doing extensive work with sales and marketing functions, Dr. Harrington began to realize that the concept of lost-opportunity cost also had a major impact on the corporate bottom-line. As a result, lost-opportunity cost was added as a new element of indirect poor-quality cost.

Today, we not only recognize the measurability of quality costs but that these costs are central to the management and engineering of modern total quality control as well as to the business strategy planning of companies and plants.

Dark Side of Quality Cost Analysis

Quality Cost Analysis looks at the company's costs, not the customer's costs. The manufacturer and seller are definitely not the only people who suffer quality-related costs. The customer suffers quality-related costs too. If a manufacturer sells a bad product, the customer faces significant expenses in dealing with that bad product. Most software projects involve conscious tradeoffs among several factors, including cost, time to completion, richness of the feature set, and reliability. There is nothing wrong with doing this type of business tradeoff, consciously and explicitly, unless you fail to take into account the fact that some of the problems that you leave in the product might cost your customers much, much more than they cost your company.

The external failure costs are borne by customers, rather than by the company. These are the types of costs absorbed by the customer who buys a defective product as follows: wasted time, lost data, lost business, embarrassment, demos or presentations to potential customers fail because of the software, failure when attempting other tasks that can only be done once, cost of replacing product, cost of reconfiguring the system, cost of recovery software, cost of technical support, and injury/death.

These are the types of costs absorbed by the seller that releases a defective product as follows: technical support calls, preparation of support answer books, investigation of customer complaints, refunds and recalls, shipping of updated product, added expense of supporting multiple versions of the product in the field, lost sales, lost customer goodwill, discounts to resellers to encourage them to keep selling the product, warranty costs, liability costs, government investigations, penalties, and all other costs imposed by law.

The problem of cost-of-quality analysis is that it sets us up to underestimate our litigation and customer dissatisfaction risks. When the total cost of quality associated with a project is estimated, a fairly complete analysis has been done. But if customers' external failure costs are not taken into account at some point, huge increased costs over decisions would be surprising.

Quality System Inspection Technique

The Quality System Inspection Technique (QSIT)[3] process encompasses three new approaches that are different from the traditional quality system inspection in recent years.

The first is the concept of a "top-down" process. With traditional inspections—and for those that will continue in the future under the "for-cause" approach—the investigator begins the process by looking at raw data or records that relate to device problems and production issues. These "bottom-up" inspections work their way up the system, with the goal of reaching the root cause of the problem(s) and working out a solution with the firm. As more information is needed, the investigator bores down farther into the records, taking a top-down approach.

The second new approach is the concept of record sampling. Sampling allows the investigator to look at a selection of records by using statistical tables as a guide for determining how many documents to examine. This helps to keep the process moving, and to cover more ground during the inspection.

The final new approach is the concept of pre-inspection record review. With QSIT, investigators are encouraged to ask for records before the inspection actually begins. While firms are not required to send any records prior to the actual start of the inspection, data from the QSIT study indicate that this practice saved time and enhanced the efficiency of the inspection.

QSIT Principles

QSIT is an inspection process based on the subsystems of the quality system. It has been suggested that a quality system can be broken down into seven subsystems. To improve the efficiency of the inspection process, the QSIT approach focuses on the four primary subsystems: management controls, design controls, corrective and preventive actions, and production and process controls. The other three subsystems—material controls; records, documents, and change controls; and equipment and facilities controls—are addressed by QSIT through linkages rather than through specific coverage. As indicated in the guide, linkages allow the investigator to leave a subsystem in order to cover another aspect of the quality system. They are used because subsystem probes cannot be performed exclusive of other aspects of the system. The QSIT approach reduced the time required for comprehensive inspections and appeared to focus the inspections on the key areas of company quality systems.

Two aspects of QSIT distinguish it as a unique inspection process: the principles of management responsibility and of Corrective And Preventive Actions (CAPA). The principle of management responsibility means that QSIT places a major emphasis on evaluating a firm's management and how well it is doing in establishing an effective quality system. QSIT inspections both start and finish with a re-

view of the firm's management controls. The QSIT approach actually advises the investigator to consider the fact that product, design, and CAPA problems found during the inspection can be used to tie back to the inability of the firm's management to establish an effective quality system.

The principle of CAPA involves a new way for the investigator to look at the firm's systems for handling CAPA information. The investigator is aware that, under QSIT, it is possible for product and process non-conformances to occur, just as they would under any system. But what is more important is the firm's process for handling nonconformance data: the investigators are told that the system for handling quality data is more important than the actual events that reside in the data. Because the investigators are checking the system, QSIT will result in more conclusions being made about the system. It thus becomes especially beneficial for companies to establish and maintain good systems prior to inspection. A benefit of the QSIT approach is that it should result in the dialogues between the agency and the firms being on systems issues, rather than on issues of product and process problems.

Notes

1. Total Quality Management (TQM)　全面质量管理
2. Appraisal costs　评价成本
3. Quality System Inspection Technique (QSIT)　质量系统检测技术

Exercises

1. What does Total Quality Management mean?
2. What does quality cost consist of?
3. How to think of the cost of customer?
4. What is the function of QSIT?

Chapter 9 Management Information Systems

Unit 1 Introduction to Management Information Systems

Definition of Management Information Systems

Managers have always used information to perform their tasks, so the subject of management information is nothing new. What is new is the ease with which accurate and current information can be obtained and communicated. The innovation that makes this capability possible is computers. The first application of computers as an information system was called the Management Information System (MIS).

Management information systems can be defined as a combination of computers and people that are used to provide information to aid making decisions and managing a firm. [1] The MIS is one of the major computer-based information systems. Its purpose is to meet the general information needs of all the managers in the firm or in some organizational subdivision of the firm. The MIS provides information to users in the form of reports and outputs from simulations by mathematical models. The report and model output can be provided in a tabular or graphic form.

Essential Components of MIS

Namely, there are three essential components for the MIS. They are management, information and system.

What Is Management?

Traditionally, the term "management" refers to the activities (and often the group of people) involved in the four general functions: planning, organizing, leading and controlling of limited resources. Management is creative problem solving. This creative problem solving is accomplished through four functions of management. The intended result is the use of an organization's resources in the way that accomplishes its mission and objectives. Note that the four functions recur throughout the

organization and are highly integrated. Organizing is divided into organizing and staffing so that the importance of staffing in small businesses receives emphasis alongside organizing. In the management literature, directing and leading are used interchangeably.

Planning is the ongoing process of developing the business' mission and objectives and of determining how they will be accomplished. Planning also includes the broadest view of the organization, e. g. , its mission, and the narrowest, e. g. , a tactic for accomplishing a specific goal.

Organizing is to establish the internal organizational structure of an organization. It focuses on division, coordination, and control of tasks and the flow of information within the organization. This function makes managers distribute authority to job holders.

Leading is to influence people's behavior through motivation, communication, group dynamics, leadership and discipline. The purpose of directing is to channel the behavior of all personnel to accomplish the organization's mission and objectives while simultaneously help them accomplish their own career objectives.

Controlling is a four-step process of establishing performance standards based on the firm's objectives, measuring and reporting actual performance, comparing the results of the real performance and the standard performance, and taking corrective or preventive actions as necessary.

Each of these functions involves creative problem solving. Creative problem solving is broader than problem finding, choice making or decision making. It extends from analysis of the environment within which the business is functioning to evaluation of the outcomes from the alternative implemented.

What Is Information?

We define information as some tangible or intangible entity that reduces uncertainty about some state or event. As an example, consider a weather forecast predicting clear and sunny skies tomorrow. This information reduces our uncertainty about whether an event such as a baseball game will be held. Information that a bank has just approved a loan for our firm reduces our uncertainty about whether we shall be in a state of solvency or bankruptcy next month. Information derived from processing transactions reduces uncertainty about a firm's order backlog or financial position. Information used primarily for control in the organization reduces uncertainty about whether the firm is performing according to plan and budget.

Another definition of information has been suggested: "Information is the data that has been processed into a form that is meaningful to the recipient and is of real perceived value in current or prospective decisions". This definition of information stresses the fact that data must be processed in some way to produce information. Information is more than raw data.

What Is a System?

A system is a group of elements that are integrated with the common purpose of achieving an objective. An organization such as a firm or a business area fits this definition. They work on achieving particular objectives that are specified by the owners or management.

There are five basic elements in a system (Figure 9-1). They are input, transformation, output, control mechanism and objectives. Resources flow from input to output through the transformation element. The control mechanism monitors the transformation process to ensure that the system meets its objectives.

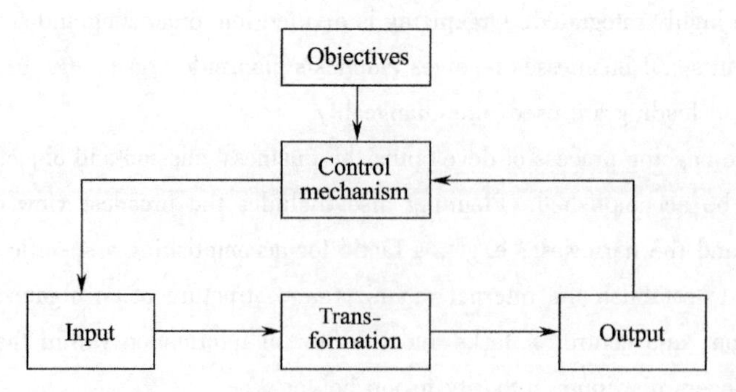

Figure 9-1 The Basic Elements in a System

If a system exits on more than one level, it can be composed of subsystems. An automobile is a system composed of subsidiary systems such as the engine system, body system and frame system. Each of these systems is composed of lower-level systems. For example, the engine system is a combination of a carburetor system, a generator system, a fuel system, and so on. These systems may be subdivided into still lower-level systems or elemental parts. The parts of a system, therefore, may be either lower-level systems or elemental parts. On the other hand, when a system is a part of a larger system, the larger system is the super-system. For example, a town's government is a system, but it is also part of a large system—the government of a state or province. The state or provincial government is a super-system of the town government and is also a subsystem of the national government.

Functionalities and Architectures of MIS

The MIS is a tool with IT for the execution of the management plan. The followings are the major functions of MIS.

Data input and maintenance.

Monitoring.

Evaluation.

Information dissemination to the public.

Provide managers with reports or with on-line access to the organization.

Serve to functions of planning, controlling, and decision making at the management level.

Summarize and report on company's basic operations.

Basic Architectures of MIS

The physical composition of MIS includes:

- Computer hardware systems;
- Computer software systems;
- Data and its storage medium;
- Communications systems;
- Information collecting and handling equipment of non-computer systems;
- Rules and regulations;
- Staff.

The function structure of management information systems are:

- Information collection;
- Information storage;
- Problem solving;
- Converses and the information output;
- Information management organization.

A management information system can be divided into the following four levels:

- Business processing;
- Service information processing;
- Tactic information processing;
- Strategy information processing.

The management information system can be divided into several interdependence subsystems according to management functions:

- Marketing subsystem;
- Production subsystem;
- Rear service subsystem;
- Human affairs subsystem;
- Financial subsystem;
- Information management subsystem;
- High level management subsystem.

Development Approaches to MIS

There are several approaches to develop an MIS. Among them, life cycle, prototype and object-oriented development approaches are often used.

Life Cycle Development Approach

The life-cycle concept is a "cradle to grave" approach to thinking about products, processes and services. System Development Life Cycle (SDLC) is the overall process of developing information systems through a multi-step process from investigation of initial requirements to analysis, design, implementation and maintenance.[2] There are many different models and methodologies, but each generally consists of a series of defined steps or stages.

Various SDLC methodologies have been developed to guide the processes involved, including the waterfall model (which was the original SDLC method), Rapid Application Development (RAD), Joint Application Development (JAD), the fountain model, the spiral model, build and fix, and synchronize-and-stabilize. Frequently, several models are combined into some sort of hybrid methodology. Some methods work better for specific types of projects, but in the final analysis, the most important factor for the success of a project may be how closely the particular plan was followed.

In general, an SDLC methodology follows the following steps:

The existing system is evaluated. Deficiencies are identified. This can be done by interviewing users of the system and consulting with support personnel.

The new system requirements are defined. In particular, the deficiencies in the existing system must be addressed with specific proposals for improvement.

The proposed system is designed. Plans are laid out concerning the physical construction, hardware, operating systems, programming, communications, and security issues.

The new system is developed. The new components and programs must be obtained and installed. Users of the system must be trained in its use, and all aspects of performance must be tested. If necessary, adjustments must be made in this stage.

The system is put into use. This can be done in various ways. The new system can phased in, according to application or location, and the old system is gradually replaced. In some cases, it may be more cost-effective to shut down the old system and implement the new system all at once.

Once the new system is up and running for a while, it should be exhaustively evaluated. Maintenance must be kept up rigorously at all times. Users of the system should be kept up-to-date concerning the latest modifications and procedures.

Prototyping Approach

The traditional approach, however, does not provide the end user with a clear picture of the final application during the design process. The end user may be disappointed when the application does not meet expectations. This can result in many changes in design and coding and costly production delays. Some definitions about prototyping approach are listed in Table 9-1.

Table 9-1 Definitions of Prototyping Approach

Prototype	An interim version or mock-up of the application used to communicate to the end user an idea of what the final application will look like
Prototyping	The process of making a prototype
Prototyping tool	A tool that gives the end user means to visualize the final application in its various stages of development
Prototyping methodologies	The technical methods used to do prototyping
Cycle time	The time it takes the designer to implement a change in an application after the change has been requested by the end user

The prototyping approach allows the designer to use a tool, such as a mock-up. This mock-up is used early in the design process to discuss an application with an end user. The prototyping tool:

- Has a very short cycle time;
- Rapidly gives the end user a deliverable item;
- Allows the designer to let the user SEE something or FEEL something extremely early in the design process;
- Facilitates communication between the designer and the end user because what the user sees is exactly what results;
- Allows changes to be made easily and rapidly;
- Has crisp code;
- Quickly pinpoints bugs;
- Provides instant compiles and link cycles;

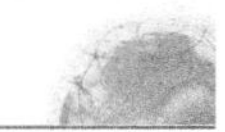

- Avoids scheduling delays;
- Results in reduced development time with happy users and happy programmers.

The prototyping course usually consists of the following steps:

Interview the End User

End users have difficulty explaining to programmers what they want and programmers have difficulty understanding what end users want. End users tend to think of things in terms of file cabinets, drawers and sheets of paper. The programmer is thinking about something totally different. Both of them are thinking about that they are going to fit something on the first screen. This causes complications later on. The interview helps determine what the end user wants to accomplish.

Design a Prototype of the Application

What you see is what you get. When designing the application, the designer uses a prototyping tool, such as a report or graph, etc., while interviewing the end user. Most users don't care what the daily input screen looks like or the processing information looks like, they care what the output of the system is. So that the end user has a clear picture of the output, the designer creates, report layouts and mock test data.

Continuous Communication Between Designer and End User

The prototyping tool provides concrete means by which the designer and the end user can discuss the design of the application. When there is a meeting between an end user and a programmer, the end user thinks that the programmer knows what is going on. And the programmer thinks the end user knows what is going on. THEY ARE PROBABLY BOTH WRONG! The end user is thinking in terms of sheets of paper, and the programmer is thinking in terms of screens and layout forms.

Words are not enough. End users' describing to the programmer what they see in their head is not enough. The prototyping tool allows the user to sit with the programmer and show them exactly what they want the application to accomplish. For example, when the end user sees a prototype report, he might say, "take this column out, add this description," etc. The design, interview, signoff can be done all in the same session, with the designer sitting down at a terminal with the end user. This can be done in an office with a screen and a printer. The report is printed out, the end user looks at it, and says, "Yes, this report looks pretty nice. This is what I want to do."

Make Changes in Prototype

A prototyping tool also results in a faster cycle time. Typical prototyping tools have a cycle time of LESS THAN TWO MINUTES. This means that a fairly significant change to the prototype can be made. Within TWO MINUTES, all the reports and screens have been adjusted, instead of waiting days as when writing the application using traditional coding methods. The user might ask for some changes, e.g., "Could you make the phone number double high so it would be easier to see on the screen?" The programmer says, "Sure," and then shows it to the user on the screen. The user might then say "It looks too thin, could you widen it?" Then when the programmer does that, the user might see that it doesn't fit on the screen, so they want the phone number back to the original way! That's what communication is about.

Sign Off on Each Phase of Prototype

The user's approval is indicated by means of a written sign off on the prototype before the coding starts. Sign off means that the end user says, "Yes, this is what I want." A sign-off phase causes the end users to share the responsibility for the completion of a phase of the application. This sign off indicates that Phase Ⅰ has been completed and Phase Ⅱ has started. Sign off trains users to agree to something—they never had to do this before. It works out well to have this agreement.

Coding

Pre-coding is coding that is done to produce the prototype or interim version of the final application. There should be a pre-coding phase so that what the end user sees in the prototype is close to the final design of the application. Pre-coding takes place before any time-consuming hard-core coding is done. Prototypes can be very, very slow in handling massive amounts of data. End users tend to become impatient when they are slow.

Testing

It is essential to test the application before giving it to the end user. For example, a company in San Diego does a lot of testing for the military. They had a system out on a ship that would crash at 2:00 a.m. every morning. What was happening was that on one person's shift, he would prop his feet up on the keyboard, not realizing that his feet were hitting a key, and at the same time was pushing other buttons. No one had ever tried that before!!

So, the testing phase is really important!

Final Sign Off

This is the final sign off on the REAL application.

Object-Oriented Development Approach

An upsurge of object-oriented programming (OOP) method has been raised in 1990s. It is called a revolution of software exploitation method. Before OOP, programming methods are all process oriented. It is designed to solve a certain problem. The key of OOP is to divide application procedure into classes. Classes can be inherited, which enable the procedure to have the reusability.[3]

Object

Object is the basic element of software system. It is a kind of thing. It may be real, like equipment. It also may be abstract, like an ordering process. These objects already contain the construction of data; also include the behavior (namely operation).

Class

Class is abstracted from objects that have consistent data construction and behaviors. Each class has innumerable objects, and each object is a real example of corresponding class.

Inherit

Inherit is the relations between classes. The existing class is called father class. The new class inherited from father class is called sub-class. The sub-class can share the attribute and operation of father class.

Object-oriented modeling technology OMT (Object Modeling Technique) uses three kinds of models:

(1) Object model. Descript object structure of system, including relations between objects, object attribute and operation. It is expressed by object chart.

(2) Dynamic model. Descript the object condition of class and correct order of events. It is ex-

pressed by the state diagram.

(3) Function model. Only consider what system do, and don't care how to do. Its description tool is data flow chart.

Notes

1. Management information systems can be defined as a combination of computers and people that are used to provide information to aid making decisions and managing a firm.

句意：管理信息系统可以定义为计算机和人的结合，它被用来为帮助公司制定决策和管理提供信息。

2. System Development Life Cycle (SDLC) is the overall process of developing information systems through a multi-step process from investigation of initial requirements to analysis, design, implementation and maintenance.

句意：系统开发生命周期是通过一个从初始需求的调研到分析、设计、实施以及维护的多步骤的开发信息系统的全过程。

3. The key of OOP is to divide application procedure into classes. Classes can be inherited, which enable the procedure to have the reusability.

句意：面向对象的方法的关键就是把应用程序划分为类。类可以被继承，这就使得程序具有重复利用性。

Exercises

1. What are the three essential components of MIS?
2. What are the major functions of MIS?
3. How many steps does an SDLC methodology follow? What are they?
4. What are the advantages of the prototyping approach?
5. What are the relations among object, class and inherit?

Unit 2 Systems Analysis

Overview of Systems Analysis

Systems analysis is a relatively new field in mankind's knowledge but demands for systems analysis have existed for many centuries before the introduction of computers. In the mid 19 century, practitioners in labor, organization and methodology had established many improved methods of working. This is the first approach to systems analysis.

With the development of information technology, systems analysis science also develops more and more vigorously and has a significant role in a life cycle of an information technology application and of information technology projects in general.

Systems analysis is an explicit formal inquiry carried out to help someone (referred to as the decision maker) identify a better course of action and make a better decision than he might otherwise have made.[1] Systems analysis usually has some combination of the following: identification and re-identification of objectives, constraints, and alternative courses of action; examination of the probable consequences of the alternatives in terms of costs, benefits, and risks; presentation of the results in a comparative framework so that the decision maker can make an informed choice among the alternatives.

In simple words, systems analysis is the requirement and request to analysis the users and the markets.

Importance of Analysis in System's Life Cycle

The primary purpose of the analysis activity is to transform its two major inputs, user's policy and systems charter, into structured specification. This involves modeling the user's environment with functioning diagrams, data flow diagrams, entity-relationship diagrams and the other tools of the systems analysis.

The quality of systems analysis can have a big effect on speed of systems designing, the programming and time for testing because 50 percent of faults in the system originated from shortcomings in systems analysis. Programmers can complain about slow speed of work because they have to review analysis and designing. This shows the bad quality of systems analysts (because of inadequate experience or improper working attitude). Moreover, speed of systems analysis activities is also a very important issue because there are always complaints about time. And the products of the systems analysts are often specification description and diagrams of technical nature and these are not highly valued. For users, what they care is what functions the program can perform if it meets the professional need dictated by the system; if its reliability is proved while testing with real figures, if the interface is users-friendly. Analysis plays a very important part in the life cycle of the system because this activity relates to almost all other activities and all subjects participated in the system.

Goals and Functions of Systems Analysis

It is a very difficult task to determine the demands of the users and the market. On the one hand, because of lack of the computer knowledge, it is unable for the users to determine what the computer actually can do or can't do at first. Therefore they cannot accurately express their own demands all of a sudden. Sometimes the demands they proposed are often uncertain. On the other hand, the designers lack the specialized knowledge of the users and some concrete market domain. It is not easy to understand the true demands of the users and market. Sometimes even the demands of the market and the users have been misunderstood. In addition, the appearance of new hardware and new software technologies also can cause the changes of demands of the users and market. Therefore the designer should carry on the exchange with the user thoroughly and unceasingly, and conduct the thorough investigation and study to the market unceasingly to determine the actual demands of users and the market gradually.

Systems analysis is the main and essential stage of management information systems development and the essential character who is responsible for this stage is the systems analyst. The key question to

complete this stage task is the communication between the development personnel and the users. Because the users must inspect the accuracy and integrity of systems analysis documents, but the systems estimator then carries on the new system design according to systems analysis documents. The users often only have management and service knowledge, but the systems estimator often only has the computer knowledge. There is an obvious gap between them. Therefore the systems analyst is the bridge between the systems users and the systems development personnel.

Therefore, in the very great degree, the systems analysis is said to be a communication process. It is an understanding and the explanation process. The systems analysis is also called the logical design. It is an important stage in systems development life cycle. Then, why to or why not to carry on the systems design compile the procedure directly with programming tools? Some people even believe the designer may develop the high grade management information systems so long as they grasp several machine languages skillfully. In fact, the programming is only a small part of work in a management information system performance history. The massive practices indicate that the developed information system is due to fail if the designer considers how to do directly before he makes clear about what to do. So the basic task and goal of systems analysis is to make clear about what to do.

Functions of Systems Analysis

The systems analysis is the main stage in a management information system construction. The systems analysis completed in this stage has three functions.

- It is the logical designing of a management information system;
- It is the foundation of the systems design, systems implementation and systems test;
- It is the basis of systems approval.

Importance of Requirement Analysis

Requirement analysis is the beginning of management information systems development. Whether its results have reflected the users' actual requests accurately will directly affect the following each stage. And it will also affect whether the design result is reasonable and practical.

The duties of requirement design are to investigate carefully and closely the object to be processed in real world, to understand the working conditions of the present system and to identify all the requirements of both market and users. Then the functions of the new system are determined. The new system must consider the next possible expansion, change and the development. The management information system should not be designed and developed only according to the current applications.

The key point of requirement analysis is the investigation, collection and analysis of the information requirement, the processing request, secure and complete request, the management decision-making request and requests of the market and users in management information systems.[2] The information requirement refers to the content and the nature of the information obtained from MIS in the market and users requirements. The data request, that is, which data should be stored in MIS, can be derived from MIS according to the user and market requirements. The processing request refers to what processing function the user and the market require, what is the requirement for the response time, the processing way is the batch processing or on-line processing and so on. The new system functions must be able to satisfy the user's and the market's information requirement, processing request, secure and complete request, the management decision-making request and so on.

Contents and Principles of Systems Analysis

Analysis is the focus of systems developing and is also the stage when systems designers have to work at two levels of definition regarding the study of situational issues and possible solutions in terms of "what to do" and "how to do".

Systems analysis process in its most general form includes the following main contents:

- Identify the operations of the existing system;
- Understand what the existing system is doing;
- Understand the need of the users;
- Decide on what the new system should do;
- Decide on how the new system will function.

At the moment, there is no method that ensures success and that can be viewed as a "right" way for analysis but the application of the structured systems analysis increases the chance of success for most of typical applications and it proves efficient in a range of analysis in real life.[3]

Until today, systems approach is still viewed as a sound foundation for the structured systems analysis. The structural systems analysis is a modern approach to different analysis and design phases of the systems development process which is accepted because of its advantages over other traditional approaches. The structural systems analysis has the following main characteristics:

(1) The system is developed in the top-down order.

(2) During systems analysis and design, several tools, techniques and models are used to record and analyze the current system and new requirements of users, and define a format for the future system.

(3) The major tools used in the structural systems analysis include: function diagram, data flow diagram, data dictionary, process specification, Entity-Relationship diagram.

(4) Separation between physical model and logical model. A physical model is often used in surveying the current system and designing the new system while a logical model is used in analyzing system's requirements. This is a significant advantage brought about by the structural systems analysis method.

(5) Acknowledging users' role in different steps of systems development.

(6) Different steps in the structural analysis and designing can be carried on at the same time rather than in one by one order. Each step can improve the analysis and designing made in a previous step.

(7) Structural analysis is supported by advanced technology in both hardware and software, therefore systems development with this method is less complicated.

(8) Structural analysis combined with the prototype method can help users and analysts have an idea of the new system and help make best use of both methods.

Two Models of the Structural Systems Analysis

Waterfall Model

Waterfall model has been a basis for a majority of the structural systems analysis methods since the 1970s. This model consists of different phases carried out one after the other. Each phase can be assigned to a group of experts.

Spiral Model

Spiral model was initiated by Barry Boehm, has become more and more popular and a basis to iterative development systems. This approach also consists of various phases carried out one after the other as in the Waterfall model. However, the systems development process is divided into different steps and smoothened after repetitive steps, the system becomes more perfect after each repetitive steps. In this model, systems developer can hand system's functions over to users more frequently rather than giving them all the functions at the end of the development process.

Popular Methods of Systems Analysis

In the following contents, we will be introduced how to use a range of tools and techniques that enable analysts to have a better understanding and to find a solution for every single case of applications. The usage of tools and techniques forms an approach named the structured systems analysis. The structured systems analysis originated from observations that principles of structured programming can also be applied to analysis and designing stage.

In the process of systems analysis, analysts often construct models to give an overview or stress on aspects of the whole system. This enables analyst to contact users in the best way and when users'need is changed, it is possible to modify or construct a new model. Analysts use models to:

- Concentrate on important features of the system, pay less attention to less important ones;
- Be able to respond to changes or changes in user's requirements with low cost and risk;
- Properly understand users' environment and write documents in the same way that designers and programmers construct the system.

Several important modeling tools used in systems analysis are: functional diagram, data flow diagram, the data dictionary and the process specification.

Functional Diagram (FD)

A functional diagram is used to show system's functions that will be constructed and the implementation process of data diagram. Moreover, function diagram will also be used to determine the appearance frequency of smaller process in the data flow chart. If during the construction of functional diagram, analysts identify new functions, the analysts need to determine whether it is a wrong move to ignore the discovered functions. It is necessary to decide to add or remove in the most appropriate way. Functional analysis with the help of modeling tools provides important details that will be used often in later stages of analysis. With detailed job description, information processing and exchanging process, input and output of every function will help analysts understand more clearly system's requirements. However, it is necessary to note that the function approach is not a comprehensive approach. A function diagram only shows what to do instead of how to do. In a functional diagram, a function is divided into many smaller functions and each smaller function contains many even smaller ones. Constructing diagram is a process of division, from a higher function to appropriate smaller functions. Diagrams need to be presented clearly, simply, exactly, fully and evenly.

Data Flow Diagram (DFD): describe the information flow in the system

Data flow diagram includes four main purposes:

(1) Analysis: DFD is used to determine requirements of users.

(2) Design: DFD is used to map out plan and illustrate solutions to analysts and users while designing a new system.

(3) Communication: One of the strength of DFD is its simplicity and ease to understand to analysts and users.

(4) Documents: DFD is used to provide special description of requirements and systems design. DFD provide an overview of key functional components of the system but it does not provide any detail on these components. We have to use other tools like data dictionary, process specification to get an idea of which information will be exchanged and how.

Data flow diagram can be described in the following ways:

- What functions should the system perform?
- Interaction between functions?
- What does the system have to transfer to?
- What inputs are transferred to what outputs?
- What type of work does the system do?
- Where does the system get information from to work?
- And where does it give work results to?

Regardless of the ways it is described, the data flow diagram needs to meet the following requirements:

- Without explanation in words, the diagram can still tell the system's functions and its information flowing process. Moreover, it must be really simple for users and systems analysts to understand.
- The diagram must be evenly laid out in one page (for small systems) and in every single page showing system's functions of the same level (for larger systems).
- It is the best for the diagram to be laid out with computer supporting tools, because in that way the diagram will be consistent and standardized. Also, the adjustment process (when needed) will be done quickly and easily.

The main components of data flow diagram are:

- The process: The process shows a part of the system that transforms inputs into outputs; that is, it shows how one or more inputs are changed into outputs. Generally, the process is represented graphically as a circle or rectangle with rounded edges. The process name will describe what the process does.

- The flow: The flow is used to describe the movement of information from one part of the system to another. Thus, the flow represents data in motion, whereas the stores represent data at rest. A flow is represented graphically by an arrow into or out of a process.
- The store: the store is used to model a collection of data packets at rest. A store is represented graphically by two parallel lines. The name of a store identified the store is the plural of the name of the packets that are carried by flows into and out of the store.
- External factors: External factors can be a person, a group of persons or an organization that are not under the studying field of the system (they can stay in or out of the organization), but has certain contact with the system. The presence of these factors on the diagram shows the limit of the system and identifies the system relationship to the outside world. External factors are important components crucial to the survival of every system, because they are sources of information for the systems and are where system products are transferred to. An external factor tends to be represented by a rectangle, one shorter edge of which is omitted while the other is drawn by a duplicated line.
- Internal factors: While the external factors'names are always nouns showing a department or an organization, internal factors'names are expressed by verbs or modifiers. Internal factors are systems'functions or process. To distinguish itself from external factors, an internal factor is represented by an rectangle, one shorter edge of which is omitted while the other is drawn by a single line.

You can construct DFD model of a system with the following guidelines:

- Choose meaningful names for processes, flows, stores, and terminators;
- Number of processes;
- Re-draw the DFD many times;
- Avoid overly complex DFD;
- Make sure the DFD is consistent internally and with any associated DFD.

To recap, DFD is one of the most important tools in a structured system analysis. It presents a method of establishing relationship between functions or processes of the system with information it uses. DFD is a key component of the system requirement specification, because it determines what information is needed for the process before it is implemented. Many systems analysts reckon that DFD is all they need to know about structured analysis. On the one hand, this is because DFD is the only thing that a systems analyst remembers after reading a book focusing on DFD or after a course in structured analysis. On the other hand, without the additional modeling tools such as Data Dictionary, Process Specification, DFD not only can't show all the necessary details, but also becomes meaningless and useless.

Figure 9-2 shows the data flow diagram of an airplane ticket ordering system. The function it reflects is: The travel agency inputs the passenger information (name, age, ID card number, travel time, destination and so on) into the airplane ticket ordering system. The system arranges the scheduled flight for the passenger, and prints the notice. The passenger pays for the tickets one day before with the notice. If the system testifies it is unmistakable, the ticket will be printed for the passenger.

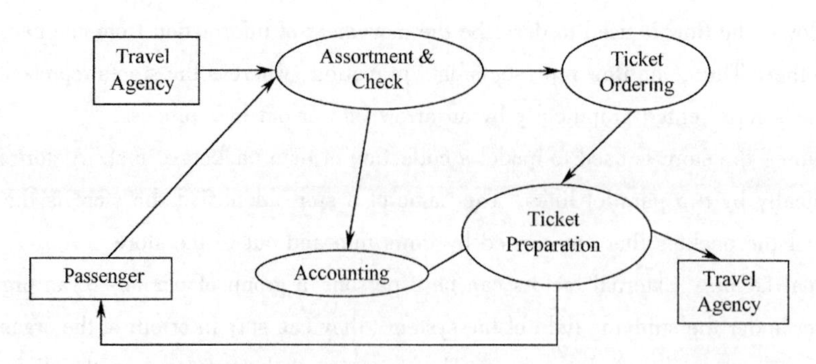

Figure 9-2 The Data Flow Diagram of an Airplane Ticket Ordering System

Data Dictionary

Data dictionary is an organized listing of all the data elements pertinent to the system, with precise, rigorous definitions so that both users and systems analysts will have a common understanding of all inputs, outputs, components of stores, and intermediate calculations. The data dictionary defines the data elements by doing the followings:

- Describing the meaning of the flows and stores shown in the data flow diagrams;
- Describing the composition of aggregate packets of data moving along the flow;
- Describing the composition of packets of data in stores;
- Specifying the relevant values and units of elementary chunks of information in the data flows and data stores;
- Describing the details of relationships among stores that are highlighted in an entity-relationship diagram.

The system analysis can ensure that the dictionary is complete, consistent, and non-contradictory. He can examine the dictionary on his own and ask the following questions:

- Has every flow on the data flow diagram been defined in the data dictionary?
- Have all the components of composite data elements been defined?
- Has any data element been defined more than once?
- Has the correct notation been used for all data dictionary definition?
- Are there any data elements in the data dictionary that are not referenced in the functioning diagrams, data flow diagrams, or entity-relationship diagrams?

Building a data dictionary is one of the relatively important aspects and time consuming of systems analysis. But, without a formal dictionary that defines the meaning of all the terms, there can be no hope for precision.

An Example of Data Dictionary

The real estate management system implements management with the computer to assignment, adjustment the real estate, or to calculate the house rent. The inhabitant may inquire the housing situation and the amount of house rent, while the management department of real estate may inquire or statistic real estate service condition. The system outputs the inquiry result with the screen or the statistical table.

The system has 23 data flows, which are listed in L01 ~ L23, 14 processes, which are listed in P01 ~ P14, 3 files listed in F1 ~ F3, 2 external items listed in W1 and W2, and 3 temporary data

flows. Its date dictionary is compiled in the following stages (Table 9-2 ~ Table 9-6):

Defining Data Flow

Table 9-2 The Definitions of Data Flow

Number	*Data flow name*	*Elements*	*Flow amount*	*Remarks*
L01	Management requirements	L14	L15	L16		
L02	Statistical table	{E01 + E02 + E03 + E04 + E05}		
L03	Inhabitant requests	L09	L10	L11	L14	L15		
L04	Processing and inquiry result	L07	L08		
L05	Legitimate inhabitant requests	L09	L10	L11	L14	L15		
L06	Legitimate consultation requests	L14	L15	L16		
L07	Processing results	L12	L13		
L08	Housing situation	E01 + E02 + E03 + E04 + E05		
L09	House allocation application	E05 + E06 + E07 + E08 + E09 + E10		
L10	House transposition application	E05 + E06 + E07 + E02 + E2 + E09 + E10		
L11	House returning application	E05 + E06 + E01 + E10		
L12	House allocation notice	E01 + E05 + E10		
L13	House returning notice	E01 + E05 + E10		
L14	Housing situation consultation	E05		
L15	House rent consultation	E05	E01		
L16	Statistical requests	E11		
L17	Housing records	E01 + E02 + E03 + E04 + E05		
L18	Rent	E01 + E04		
L19	Detailed list of rent specifications	{E01 + E04 + E05}		
L20	Approved allocation application	L09		
L21	House allocation list approved	E01 + E05 + E10		
L22	Transposition application	L10		
L23	House returning list	E01 + E05 + E10		

Defining Data Elements

Table 9-3 The Definitions of Data Elements

Number	*Name of elements*	*Inner name*	*Range*	*Level*	*Class/length*	*Remarks*
E01	House number	ZFBH	9000		N/4	
E02	House level	FWDJ	1 ~ 3	3 levels	N/1	
E03	House condition	FWZK	1 ~ 3	3 levels	N/1	
E04	Rent	ZJ	1 ~ 5		N/4/2	Price/㎡
E05	Inhabitant number	ZHBH	9900		N/4	
E06	Inhabitant name	ZHXM			C/8	
E07	Housing standard	ZFBZ	1 ~ 3	3 levels	N/1	
E08	Level of house on hand	XFDJ	1 ~ 3	3 levels	N/1	
E09	Level of house required	YFDJ	1 ~ 3	3 levels	N/1	
E10	Auditing	SHXM			C/8	
E11	Date	RQ			D/10	

(Contained)

Number	*Name of elements*	*Inner name*	*Range*	*Level*	*Class/length*	*Remarks*
E12	Empty house number	KFSL	9000		N/4	
E13	1 lever house number	YJSL	3000		N/4	
E14	2 lever house number	EJSL	4000		N/4	
E15	3 lever house number	SJSL	2000		N/4	

Defining Files

Table 9-4 Files Definitions

Number	*File name*	*Inner name*	*Elements*	*Index*	*Remarks*
F1	Real estate file	CKZ	{E01 + E02 + E03 + E04 + E05}	E01	
F2	Housing standard file	MMK	{E05 + E07}	E05	
F3	Rent file		{E01 + E04}	E01	

Defining Files

Table 9-5 The Definitions of External Items

Number	*Name*	*Output data flow number*	*Input data flow number*	*Remarks*
W1	Management staff	L01	L02	
W2	Inhabitant	L03	L08, L09, L11, L12	

Defining Processing

The processing logic of all process in this example are not very complex. They can be defined in structured language.

Table 9-6 Processing Definitions

Number	Name	Input Data	Output Data	Pre-process	After-process	Related Files	Processing Logic

P1	Check legitimate	L01, L03	L05, L06	empty	P2. 1,
					P3. 1
						none	IF having management right
							DO P3. 1
							ENDIF
							IF Inhabitants
							IF housing qualifications
							DO P2. 1
							ELSE
							DO P3. 1
							ENDIF
							ENDIF

P2. 2. 1	Approve housing qualifications	L09	L20	P1	P2. 2. 1
					P2. 3. 1
						F01, F02	Read F02 housing qualifications
							IF satisfied
							Do P2. 2. 2
							ELSE
							Display ‘not Satisfied’
							ENDIF

The Process Specification

As we know, there are a variety of tools that we can use to produce a process specification: decision tables, structured English, pre/post conditions, flowcharts, and so on. Most systems analysts use structured English. However, any method can be used as long as it satisfies two important requirements:

(1) The process specification must be expressed in a form that can be verified by the user and the systems analysts.

(2) The process specification must be expressed in a form that can be effectively communicated to the various audiences involved.

The process specification represents the largest amount of detailed work in building a system model. Because of the amount of work involved, you may want to consider the top-down implementation approach: begin the design and implementation phase of your project before all the process specifications have been finished.

The activity of writing process specifications regarded as a check of the data flow diagrams that have already developed. In the writing process specifications, you may discover that the process specifications needs additional functions, input data flow or output data flow... Thus, the DFD model may be changed, revisions, and corrections based on the detailed work of writing the process specifications.

Requirements Analysis Documentation

The compilation compendium of requirements analysis documentation can be divided into the following parts:

(1) Introduction.

① Compilation goal.

② Background explanation.

③ Terminology definition.

④ Reference.

(2) Task outline.

① Goal.

② User's characteristic.

③ Assumptions and restraints.

(3) Requirements stipulation.

① Stipulation to function.

② Stipulation to performance.

- Precision;
- Time response request;
- Flexibility.

③ Input/output request.

④ Data management ability request.

⑤ Breakdown processing request.

⑥ Other special request.

(4) Operating environment hypothesis.

① Equipment.

② Supporting software.

③ Interface.

④ Control.

(5) Abbreviation table.

(6) Reference.

Notes

1. Systems analysis is an explicit formal inquiry carried out to help someone (referred to as the decision maker) identify a better course of action and make a better decision than he might otherwise have made.

句意：系统分析是一种明确而且正式的调查，通过实施它来帮助某人（是指做决策者）更好地定义决策过程和有可能使决策者做出一个更好的决策。

2. The key point of requirement analysis is the investigation, collection and analysis of the information requirement, the processing request, secure and complete request, the management decision-making request and requests of the market and users in management information system.

句意：需求分析的关键是对管理信息系统中的信息要求、处理要求、安全和完备要求、决策管理要求以及市场和用户的要求的调查、收集和分析。

3. At the moment, there is no method that ensures success and that can be viewed as a "right" way for analysis but the application of the structured systems analysis increases the chance of success for most of typical applications and it proves efficient in a range of analysis in real life.

句意：目前，没有一种可以保证成功的方法，或者被认为是系统分析的最正确的方法。但是对于大多数典型的应用，结构化系统分析的应用无疑可以大大增加成功的概率。而且在实践中一定范围内的分析已经证明它是十分有效的。

Exercises

1. What is system analysis? And what are its goals and functions?

2. To analysis an information system, what principles should an analyst follow?

3. How many popular methods are there for system analysis? Give an example for each method.

4. Give an example, and try to write an outline of requirements analysis documentation.

5. Investigating library system in your university, compile its data dictionary using the techniques learned in this unit.

Unit 3 Systems Design

The Goals of Systems Design

Systems analysis gives a solution to the question about what a management information system does. It describes the goals and functions of a management information system, while systems design solves the problem of how to do for a management information system. Systems design carries on the total design and the detailed design for the management information system.

The systems design is the physics design stage of new systems. According to logical model and function requests of new systems which were determined in analysis stage, design a plan which can be implemented in the computer network environment, namely establish the physical model of new system.[1]

The emphasis of this stage is to design the overall structure of software, the function module and theirs mutually relations, provide the essential explanation for the coding and design the database. These work mainly be undertaken by the computer specialized technical personnel. It is equal to the engineering designing and blue painting drawing of a building.

This unit introduces techniques for the design of interfaces, menus, and databases, based on the requirement specification worked out during the system analysis phase (functioning diagram, relationship diagram, data flow diagram...). At the end of this phase, you need to identify the borderline between the computer system and human being and find the answer to the question about how to attain the system's objectives.

By definition, analysis and design are two separate activities. However, in practice, the two development activities are so intertwined that no one can say exactly when analysis ends and design begins. For example, design ideas are often conceived during the preparation of documents such as Functioning diagrams, Data flow diagrams, Entity relationship diagrams, Data dictionaries, Specification process from the formal specification of user requirements

The design of an appropriate information system requires that analysts understand the goals and objectives of management. They must also be sensitive to changes that may occur to these goals and objectives over time in response to shifts in the competitive environment.

The design objectives specified in the user implementation model are the quality of the design. The ability of the programmers to implement a high-quality, error-free system depends very much on the nature of the design created by the designer; also, the ability of the maintenance programmers to make changes to the system after it has been put into operation and to the expanded system depends on the quality of the design.

The Contents of Systems Design

Module Design of Subsystem Functions

The overall plan and the subsystem division in system analysis have defined the upper functions of system in fact. In order to truly realize these functions, it is often necessary to continue to decompose. Finally a multilayered function graduation structure has been obtained. The method to design a management information system into certain functions module is called modulation. This is an important thought. The module division is the same as the subsystem division. That is, they also divide a system into some parts which are small scales, relatively independent in functions and easy to modify and establish.

Code Design

Code design means carrying on the specific code design according to the code system of replacing the Chinese characters with the characters and the numerals determined in system analysis. The concrete contents include:

- Code structural design;
- The use scope, the deadline and service revision jurisdiction;
- Code table establishment.

Input Design

The accuracy of input data plays the decision role in the overall system quality. Improper input design may cause the mistake of the input design. Even extremely correct computation and processing will not definitely obtain the correct output. There is a maxim in computer profession "the input is trash, the output is also trash." Therefore, the input design must both provide the convenience interface to the user and have the strict inspection and error correction function in order to reduce the possible mistake as far as possible. At the same time, try to use the automatic input way like bar code input, the on-line monitor and so on as far as possible.

Output Design

Because output relates to the users directly, therefore the design starting point is to guarantee the output achieves the user's requests, reflects and organizes each kind of useful information correctly. At the same time, improve enterprise's report form transmission means and reduce the printing report form as far as possible by this. And enhance the working efficiency by using electronic form transmission means.

Database Design

The database design mainly includes the logical design and the physical design of the database in the system design stage. It requires synthesizing the output design and the data demand of each enterprise department. By analyzing the relations between each data, according to provided functions and description tools, design the data model with suitable scale, few data redundancy, high withdrawal efficiency, which also can reflect the data relations correctly and satisfy many kinds of inquiries request.

Systems Reliability Design

System reliability design includes the following contents:

- System continuously normal operation capability;
- Power failure emergency procedures of the system;
- The protective measures such as quakeproof, the fire protection and anti-radar strike of the system;
- Breakdown restoration of hardware and software, Data backup design, the spare equipment

preparation.

Systems Security and Secret Design

The security of management information system is an extremely important link in the system development and operation. First, make every effort to stop the information losing and being destroyed deliberately. Second, prevent the computer-related crime. The former mainly includes the operation mistakes or information loss because of other mistakes and computer virus'invasion and so on. It needs to be guaranteed in two aspects of management system and technical measures. The computer-related crime means the criminality of stealing the enterprise information through the system, non- authorized revision information, disturbance or destruction computer system's normal operation.

Interface Design with Other Systems

- On-line monitor system;
- Automatic control system;
- Joint design with external system;
- Joint design with many kinds of extraneous information networks;
- Information exchange of Unified interface.

Basic Ideas and Features of Systems Design

Traditional systems design tool is system flow chart. But the system flow chart only describes the processing logic of source program. It emphasizes the systematic realization. To be concrete, the traditional design method adopts a design strategy from the bottom upward. In other words each concrete function module is designed first, and then the designed function modules are assembled again into a software system. Obviously, this kind of design mentality has neglected the entire software overall construction.

Because the relations between each module have not been considered from the overall situation, the system designed with this method will lack of the flexibility and maintainability. And the reliability and efficiency are relative bad, which thus influence the software system quality. In order to solve this problem, the structured design concept and method have been produced.

The lofty goals of structured design are accomplished through the consistent implementation of specific philosophical views. One of these, functionality, is stressed throughout the practice of structured design. The heart of this method is the notion that a program, a group of programs, or a group of systems is nothing more than a collection of functions.

Structured design is the result of a desire on the part of its authors to formalize what they consider to be good programming practice. A beneficial effect for the software engineering is that what is easy to maintain is also relatively easy and inexpensive to construct. Thus, structured design is a software design method which closely parallels the current trend in technology toward systems which are once easy to maintain and inexpensive, or at least cost-effective, easy to construct. [2]

The designer must at first see through the programs, modules, and routines and examine relationships. One must temporarily forget about how the systems at hand might be implemented and treat it as a collection of abstractions-logical functions. This gives the software designer a maximum freedom in examining relatively system architectures.

The outstanding unique aspect of Structured Design is that it includes several different ways of evaluating a design.

Structured design provides two effective means of evaluating software design decisions. One measures the relative effect of the decision on module quality, while the other measures the relative effect of the decision on the relationship between modules.

The introduction of Structured Design can only work when it provides all concerned with a common value system, that is, a uniform view of what is or is not desirable in a system. Although the approach may differ from organization to organization, followings are some guidelines to maximize the benefit of introducing Structured Design into an organization:

- Make it clear from the start that Structured Design is not a remedy for any problems of the organization;
- Obtain support from anyone who may be affected by this change;
- Emphasize the flexibility that Structured Design provides and downplay the misconception that it is restrictive;
- Structured Design development passes through two primary phases: Logical Design development and Physical Design development;
- Logical Design transforms the dataflow diagrams developed during the Structured Analysis phase into refining the design to reflect the kinds of quality attributes that Structured Design encourages;
- Physical Design phase involves the modification of the Logical design to accommodate whatever changes necessary.

As a process, design goes through the following stages:

- Transformation of the problem model into an initial solution model;
- Embellishment of initial design to include necessary features;
- Transformation of the structured English into pseudo code;
- Evaluation and refinement of the qualities of the design;
- Revise the design to accommodate implementation constraints.

Design is an exact blueprint of what will be built and a basis for the configuration and content of that blueprint.

The Stages of Systems Design

The activities of systems design involve developing a series of models, in much the same way that the systems analyst develops models during the systems analysis phase. The most important models for the designer are the systems implementations model and the program implementation model. The systems implementations model is divided into a processor model and a task model.

The Processor Model

At this level, the systems designer decides how the essential model should be allocated to different processors and how those processors should communicate with one to another.

As processes must be assigned to appropriate hardware components, data stores must be similarly allocated. Therefore, the designer must decide whether a store will be implemented as a database on this processor or another. Since most stores are shared by many processes, the designer may also have to decide whether duplicate copies of the store need to be assigned to different processors. An important point for designer: processor to processor communication is generally very much slower than communication between processes within the same processor. Thus, the system designer will generally try to group processes and stores that have a high volume of communication within the same processor.

During designing a process, the systems designer should raise the following major issues:

Cost: Depending on the nature of the system, the systems designer must choose the most economical solution for each practical case: a single processor implementation or group of processors.

Efficiency: The systems designer is generally concerned with the response time for computer systems. Therefore, the designer must choose processor and data storage devices that are fast enough and powerful enough to meet the performance requirements specified in the user implementation model.

Security: The end user may have security requirements that dictate the placement of some processor and sensitive data in protected location.

Reliability: The end user will normally specify reliability requirements for a new system; these requirements may be expressed in terms of mean time between failures or mean time to repair, or system availability. Alternatively, the designer may decide to implement redundant copies of processes and/or data on multiple processors, perhaps even with spare processors that can take over in the event of a failure.

Political and operational constraints: The hardware configuration may also be influenced by political constraints imposed directly by the end user, by other levels of management in the organization and operating all computer system.

Task Model

When the processes and stores have been allocated to processors, the systems designer assign processes and data stores to individual propose tasks.

Program Implementation Model

At the level of an individual task, the systems designer has already accomplished two level of process and data storage allocation. Within an individual task, the computer operates in asynchronous fashion: only one activity at a time can take place. The most common model for organizing the activity within a single, synchronous unit is the structure chart, which shows the hierarchical organization of modules within one task.

Tools and Techniques of Systems Design

The design of information systems requires that systems analysts understand how information flow in an organization, how it relates to decision making, and how it contributes to organizational goals and objectives. That is why systems analysis and systems design are inextricably linked. While data are being collected on the problem environment and the analyst is learning about the information needs of the user, rough ideas about the way to improve the information function are being formulated in the analyst's mind.[3] The analyst may begin to sketch these ideas as flow charts or data flow diagrams. However, not until all data are collected and information problems are thoroughly analyzed can preliminary ideas about the design of the new system be refined. The analyst can then focus on design specifies such as input/output subsystems, processing, and data base design. Most analysts continue to use structured techniques to help them plan system components and their structure, choosing appropriate tools and techniques to aid them in the design process. The completed design should lead efficiency in satisfying users' needs and also facilitate maintenance during the life cycle of the system. Each design tool has good features and bad. Selection of design methodology will depend on the background and experience of analysts (designer), the size of the development time, resources allocated to development, the time allowance for the project, and the type of application under development. When choosing a specific tool for a specific task at hand, the following main questions should be considered:

- Will it help the team arrive at an understanding of the system under development?
- Is it easy to learn? Easy to use?
- Will it serve user-analyst communication?
- What does it cost?
- What data structures does it use?
- What data flow and control features does it have?
- Is the technique manageable?

There are many ways to approach system design and many tools and techniques that contribute to the design process. In this part we will find a discussion on other methodologies that are comely used by analysts. Many of these are complex, so what appears here is merely a brief introduction to their features, benefits and limitations. Unfortunately, the scope here limits the number of design methodologies that can be presented. In the system development process, many design methodologies already exist, but one of the challenges of systems design is to select tools and techniques that are appropriate for an application under development from the wide range of available options. The main selection that follow: ISDOS, Pseudo-code, Structured design, the Jackson Design methodology, HIPO, SADT, Walkthroughs, Entity Relationship model, Data Structure Diagrams, Semantic Data Model and CASE Method (Oracle).

Information System Design and Optimization System (ISDOS)

Information System Design and Optimization System (ISDOS) will be able to generate system specifications from user requirements recorded in a machine-readable form, design an optimal system to meet these specifications, and construct code for operational system.[4]

ISDOS begins its role in systems development once the information problem has been defined by the user and analyst and stated in PSL (Problem Statement Language), a structured format. It allows the people who define the problem (user or analyst) to state what is wanted without having to specify how these needs should be met. It is not a programming language but one that lets the user state what outputs the system should produce, what data elements the output should contain, what formula should be used to compute output values, what input the system will use, and so on. It is a language to describe systems. A computer program called PSA (Problem Statement Analyzer) then analyzes input in PSL to ensure a complete and error-free statement of the problem. A coded statement of the problem is

then sent to SODA (System Optimization and Design Algorithm), software for the physical systems design phase, which generates specifications for the actual construction of programs, for hardware, for the database, and for storage structures.

The two ISDOS modules to receive this output are the Data Re-organizer who constructs files, storing data on selected device in the form specified, and the Code Generator, which organizes the problem statements into programs. Finally, a module called the System Director produces the target information processing system, using the generated code, the stored data and the timing specifications as determined by the physical design algorithm.

ISDOS has been under development for more than a decade. Whether it will ever succeed in optimizing systems design remains uncertain. But progress is being made towards the automation of some development stages.

Pseudo-code

Pseudo-code can be used to describe an algorithm. Although pseudo-code resembles Structured English (SE) in using a restricted subset of English, it may be coded and more closely resemble a programming language. Whereas SE is a useful tool when analysts communicate with end users, pseudo-code is oriented towards analyst programmer communication, describing program logic, using program construction. In practice, there is no standard, universal pseudo-code, but all pseudo-code has three basic structures: Sequence, Selection and Iteration.

Structured Design (SD)

Nature of SD is achieved (implemented) by dividing the system in independent modules (separate pieces) that can be designed, implemented and modified with no (or little) effect on other modules of the system. Coarse program structure, based on DFD, is depicted by means of a structure chart. This structure chart, which resembles an organization chart, shows relationships between units or modules, and how modules are combined to achieve systems (organization) and design goals. This coarse program structure is then factored, or decomposed, into a fine structure which forms the basic for implementation. In order to choose among alternatives when dividing systems into modules, it is useful to evaluate the connection between them. If there are few or no connections between modules, then it is easier to understand one module without reference to others. The notion of module independence can be described in terms of coupling and cohesiveness.

Coupling is the strength of relationships between modules (the degree to which modules are interconnected with or related to one another). The stronger the coupling between modules in a system, the more difficult it is to implement and maintain the system, because a modification to one module will then necessitate careful study, as well as possible changes and modifications, to one or more other modules. In practice, this means that each module should have simple, clean interface with other modules, and that the minimum number of data elements should be shared between modules.

Cohesion is the measure of the strength among the elements in the same module (the degree to which the components of a module are necessary and sufficient to carry out one, single, well defined function). In practice, this means that the systems designer must ensure that they does not split essential processes into fragmented modules and they does not gather together unrelated processes (represented as processes on the DFD) into meaningless modules. The best modules are those that are functionally cohesive. The worst modules are those that are coincidentally cohesive. High cohesiveness occurs when all of the module parts contribute directly to the purpose or function which the module is supposed to accomplish.

Jackson Design Methodology (JDM)

The JDM is a three-step design technique. In essence, it decomposes the design process itself.

First, the problem logic is structured. Since it is Jackson who premises that a good reliable program is closely dependent on its data structure, the analyst first defines data structures. That is, the data components are identified, the relationships between these data components are defined, and then one-to-one correspondences between data structures are defined.

Next, the analyst structures program logic based on the data structures. To do so, the analyst defines the necessary processes and relationships between these processes.

Finally, the analyst defines the tasks to be performed in terms of elementary operations and assigns these operations to suitable components of the program structure.

This methodology is not a top -down method for developing computer systems. The structure charts are drawn only after action at the lowest level has been defined. The strength of the JDM is that it results in an exact reflection of data structure in the program structure. The adaptations Jackson has made to basic structure charts are simple, but yield (effectiveness) a greatly expanded expressive capacity. Also, the extension is compatible with the notion of structure programming. However, the method is not convenient for large complex systems since it may prove difficult to derive a reasonable program structure from all data structures. A higher level structuring technique in such situation may be necessary. Since Jackson system development is a very detailed and comprehensive method, well trained analysts are required to use it. When there are pressures to develop a new system quickly, the considerable time and effort required by the method are considered disadvantages. A main criticism of JDM is that it assumes that all important structures of the problem can be seen as sequential processes.

Hierarchy Plus Input, Process, and Output (HIPO)

HIPO is a graphic technique that can be used to describe a system. A series of drawings are prepared by analysts that show the function of the system starting with general overview diagrams, then proceeding to detailed diagrams of each specific function. Originally developed by IBM as a technique to document functions of programs, HIPO is today commonly used as a design tool during systems development. The package of HIPO diagrams begins with a hierarchy diagram, called a Visual Table Contents. This diagram, which is similar in tree-like structure to an organization chart, defines the basic functions to be performed by the system and decomposes those functions into sub-functions.

Preparation of HIPO diagrams begins with a series of meetings with the user about general expectations for the system and desired output, individual functions. Circumstances under which the function will be performed, data that will be acted upon, processed and expected output, relationships between those functions. After that, HIPO diagrams are generally built from the middle up and down. Once the hierarchy diagram has been drawn, analysts review the system from the top down, looking once again at the different functional levels to see if true functional decomposition exists and whether revisions to the structure of the hierarchy are necessary. The major advantage of HIPO diagrams is their simplicity, their effectiveness as a communication tool, and the ease with which analysts can learn how to diagram-

matize systems using HIPO notation.[5] The diagrams provide a structure by which the system can be understood and a visual description of input, functions to be accomplished by a program and output. The primary disadvantage is that it is impractical to draw HIPO diagrams for very large systems.

Structured Analysis and Design Technique (SADT)

SADT is a technique to develop large and complex systems. This technique helps analysts think in a structured way about system activities, data and their interrelationships. A precise notation is used to communicate analysis and design results. The method also provides for the division and coordination of effort among the development team and for planning, managing and assessing progress of the team effort. SADT analyzes a system under development from the top down, decomposing it systematically into subsystems, creating a hierarchical parent-child structure. This top to down approach resembles the decomposition of data flow diagrams. SADT is used for both systems analysis and design. Its strength is not so much the separate representation of data and activities, but the ability to diagram the relationship between the two. With SADT the system analyst and designer, management and users can review the project top to down, or focus on a single level of detail. The technique is frequently used in large and complex development project. The major disadvantage of SADT appears to be its richness. The amount of information that SADT diagrams contain may overwhelm the user and prevent their understanding of the system as whole.

Entity Relationship Model (E-R Model)

E-R Model is used in data base design helps describe how entities in an enterprise are related to one another. The model recognizes that two sets of data may have one-to-one, one-to-many or many-to-many relationships. After the model is drawn, the analyst looks for semantic synonyms (one entity referred to by two different names) or semantic homonyms (two different entities referred to by the same name) that need correction. Perhaps an entity should be named by a more generalized term and some entities can be combined into a new single entity. In making improvements to the model, the analyst is fine-tuning the logical schema of the data which will ultimately contribute to greater processing efficiency.

Data Structure Diagram (DSD)

The DSD is another diagram of pictorially representing data relationship that is used in data base design. The DSD notation expresses relationships among records—the content of the record is documented in a data dictionary. Record names appear in boxes connected by arrows to show relationships.

Semantic Data Model (SDM)

The SDM is another useful model for logical data base design and documentation. The model might be compared to pseudo-code, but instead of expressing the structure of programs, it enables an analyst to describe the structure of data. For using SDM, the analysts must learn the terminology of the model, its rules and default values. The SDM has the following main advantages:

- The model provides a vocabulary for expressing the meaning of the data. An informal description of each record is allowed and each data item within each record is defined;
- Each attribute is precisely defined, which is essential information for understanding the data and for later programming;
- SDM allows relative data definitions;

- Allowable ranges of values of data items can be identified in the model.

CASE Method

This is the structured development technique used by ORACLE consultants, and design is one of its stages. It therefore specifies the input to our design, provides a framework for its execution and defines our target output. It comprises a number of distinct stages, each having a set of tasks and deliverables. Once checked for quality and consistency, the output of one stage becomes the input to the next. Errors are thus trapped as early as possible. Throughout the development, the information gathered is expressed in the form of models. The stages of CASE Method are:

Strategy (establish direction for information system development): A strategy study is performed in order to develop an understanding of the business and to establish the scope of the system, its priorities and constraints. The resultant business model describes the information and the functions to be performed.

Analysis (establish the detailed components of the targeted business areas): Detailed work on the business model provides complete description of the required system.

Design (establish a detailed technical solution): A model for a specific implementation is developed, based on the business model. It consists of two parts: a database design to fit the information model, and application designs to fit the function model.

Build and User Documentation (create a functionally working solution and explain the solution in the user environment): The specifications produced are implemented. Documentation is developed alongside the application system.

Transition (implement the solution in the user environment): The transition to the new system is based on a plan which evolves through the preceding stages.

Production (keeps the solution working): The system goes live.

System Design Documentation

System design documentation gives an explanation to each technical design in the system construction from the point of overall system. It is the product of system design stage. It emphasizes on the explanation to the guiding ideology and the technical route and method of system design. The compilation of system design documentation will provide the essential guarantee for the following system development work from the technology and guiding ideology.

The specific request to the system design documentation is: elaborate means, method, technical standard as well as corresponding environment request in the implementation process comprehensively, accurately and clearly. Moreover the standardized question of system construction is also an important content in the system design documentation.

System design documentation mainly includes: introduction to system development project, module design explanation, code design explanation, input design explanation, output design explanation, database design explanation, network design explanation, safe security explanation, implementation plan explanation of system design.

System design documentation includes the following 10 contents:

(1) Outlines.

① System design goals.

② System design strategy.

(2) Selection of computer system.

① Computer system selection principle.

② Plan comparison.

(3) Computer system configuration.

① Hardware configuration.

- Main engine;
- External memory;
- Terminal and external instrumentation configuration;
- Other supporting facilities;
- Network architecture.

② Software configuration.

- Operating system;
- Database management system;
- Service routine;
- Language;
- Communication software, network software;
- Software development kit;
- Chinese character system.

③ Computer system geographic distribution.

④ Network agreement text.

(4) System structure.

① Structure drawing.

② Module structure drawing.

(5) Database design.

① Database overall structure.

② Database logical design.

③ Database physics design.

④ Database performance.

(6) Code design.

① Code design principle.

② Code design proposal.

(7) System failure countermeasure.

① Breakdown preventing and controlling measure.

② The system restoring method.

(8) The information preparation plan and the implement plan.

(9) The system investment plan and personnel training plan.

(10) System testing method and plan.

Notes

1. The systems design is the physics design stage of new systems. According to logical model and function requests of new systems which were determined in analysis stage, design a plan which can be implemented in the computer network environment, namely establish the physical model of new system.

句意：系统设计是新系统的物理设计阶段，它根据分析阶段所确定的新系统的逻辑模型和功能要求，设计一个能在计算机网络环境实施的规划，顾名思义，就是建立新系统的物理模型。

2. Thus, structured design is a software design method which closely parallels the current trend in technology toward systems which are once easy to maintain and inexpensive, or at least cost-effective, easy to construct.

句意：因此，结构设计是一种软件设计方法。它紧随系统技术的最新趋势，这就使得设计易于维护且廉价，或者至少是成本效率比较优，易于构建系统。

3. While data are being collected on the problem environment and the analyst is learning about the information needs of the user, rough ideas about the way to improve the information function are being formulated in the analyst's mind.

句意：当收集与问题相关的数据时，系统分析者了解用户的信息需求，这样，关于如何改善信息功能的大致想法就在设计者的头脑中产生了。

4. Information System Design and Optimization System (ISDOS) will be able to generate system specifications from user requirements recorded in a machine-readable form, design an optimal system to meet these specifications, and construct code for operational system.

句意：信息系统设计和优化系统能以一种机器易读的形式记录用户需求从而生成系统设计规范，并能设计出一种最优的系统来满足这些设计规范和构造运行系统的代码。

5. The major advantages of HIPO diagrams are their simplicity, their effectiveness as a communication tool, and the ease with which analysts can learn how to diagrammatize systems using HIPO notation.

句意：HIPO 图的主要优点就是其简化性，作为一种交流工具的高效性，以及系统分析者能很容易地学会使用 HIPO 符号来图示系统。

Exercises

1. What is the relationship between system analysis and system design?

2. What are the contents of system design?

3. To complete a system design, how many stages should the designer follow?

4. How many tools and techniques are commonly used in systems design? Give brief introductions to them.

5. What is the function of system design documentation? What should good system design documentation include?

Unit 4 Testing and Evaluating to Systems Performance

Systems Testing

A system testing is an important link in the management information systems development. It concerns the success of management information systems development. It holds a quite important status in the information systems development. Although many management methods and measures have been adopted to guarantee the product quality, it still could not avoid making mistakes. These mistakes and flaws all hide in the system. The test is a program executive process to discover the mistakes and the flaws in the system. From this point of view, the test that can find new faults is a successful test.

A system testing is a program executive process to discover the mistakes and flaws in the system. Regarding to the test goals, there are several viewpoints in the following:

- The test is the program executive process. The goal lies in the discovery of mistakes;
- A good test lies in it can discover the mistakes that have not been discovered until now;
- A successful test lies in it has discovered new mistakes that have not been discovered before.

Moreover, the test goal should be to discover each kind of latent mistakes and flaws of the system by the least time and manpower. [1] At the same time, through the test it also can be proved whether the function and performance of management information system are accordance to requirement explanation.

There are two kinds of test methods used frequently. One is the human testing that does not rely on the computer. It can search the mistakes effectively in the procedure. One or several human testing methods should be applied for an information system testing. The other is the machine test, which is also called computer aided test, namely prepare some test programs, which are operated on the computer to search the mistakes in the procedure.

The management information system is composed with computer hardware system, network system and software system. Then system test should include above tests of each part.

System testing means to validate that the system meets its specification and the objectives of its users. System testing involves testing the system as a whole. It is a critical activity in systems integration. System testing includes interface testing, stress testing and acceptance testing.

Interface Testing

Interface is an agreed mechanism for communication between different parts of the system. System interface can be classed into the following classes:

- Hardware interfaces (Involving communicating hardware units);
- Hardware/software interfaces (Involving the interaction between hardware and software);
- Software interfaces (Involving communicating software components or sub-systems);
- Human/computer interfaces (Involving the interaction of people and the system);
- Human interfaces (Involving the interactions between people in the process).

The followings give us some common interface errors:

Interface Misuse

A calling component calls another component and makes an error in its use of its interface e. g. parameters in the wrong order.

Interface Misunderstanding

A calling component embeds assumptions about the behaviors of the called components which are incorrect.

Timing Errors

The calling and called components operate at different speeds and out-of-date information is accessed.

Hardware Interface Testing

Probably the most common problem which arises during systems integration is interface problems where sub-systems and components do not interact as anticipated by their developers. [2] These may not be detectable during earlier testing phases because of interface misunderstandings or because they are a consequence of the way in which the interface is used. In many cases it is difficult to isolate problems to a specific interface level. When the hardware is integrated, it may simply fail to work properly. Interface testing usually involves the use of specialized equipment which can monitor the signal traffic between the different hardware units. Specialists must then look for anomalies in the results of the monitoring to detect interface problems.

Software Interface Testing

Systematic interface testing requires knowledge of both the system requirements and the software organization. The tester must know how sub-systems use each other's interfaces. Because of these difficulties, incremental testing is the only practical way to test sub-system interfaces. However, preparing tests for increments may be difficult as the increment may only partially implement some system requirements.

Stress Testing

Stress testing exercises the system beyond its maximum design load. The argument for stress testing is that system failures are most likely to show themselves at the extremes of the system's behaviors. When a system is overloaded, it should degrade gracefully rather than fail catastrophically. It is particularly relevant to distributed systems. As the load on the system increases, so too does the network traffic. At some stage, the network is likely to become swamped and no useful work can be done.

Acceptance Testing

Acceptance testing is the process of demonstrating to the customer that the system is acceptable. It must be based on real data drawn from customer sources. The system must process this data as required by the customer if it is to be acceptable. Generally it is carried out by customer and system developer together. It may be carried out before or after a system has been installed.

Systems Evaluating

The Targets of System Evaluating

Strictly speaking, in the whole process of management information system development, there

should be an evaluation whenever each stage or step have completed. The comprehensive evaluation should be carried after the new system has operated for a certain time. The establishment of new system has spent a massive funds, manpower and physical resource. How about the new system performance actually? Not only is it very important to the user, also it is what the system development personnel care about. Therefore the system must be evaluated. Only after carrying on the comprehensive inspection to the new system quality, can these questions be answered. It is impossible for the new system to be the acme of perfection. Through the evaluating, the problem can be found and solved. So the system can obtain essential revision and maintenance.

Any engineering project in practical application has to consider two aspects: technical and economical considerations. Either is to obtain as far as possible system functions and as high as possible system performance in the certain economic condition limit. Either is to realize the certain function and performance with fewer expenses as far as possible.

Compared to other project systems, the evaluation of management information system also has its own characteristics. A management information system has included many factors such as the information resource, technical equipment, human and environment. The system potency is displayed through the functions and the ways of information. But the information functions are displayed in the decision—making and behaviors of human beings in the certain environment, with the aid of computer technology. Therefore, the potency of management information system may be visible or invisible, may be direct or indirect, and may be fixed or changeable. Therefore the evaluating of management information system has its complexity and particularity.

The so-called systems evaluating is to estimate, inspect, test, analyze and examine the performance of management information systems. [3] It includes the comparison as well as the evaluation of systems goal realization degree with the actual index and plan index.

The followings list the goals of systems evaluation:

- Inspect whether the systems goals, functions as well as each index of information systems achieved with the design request and the user's request;
- Inspect utilization degree of each kind of resources in information systems;
- According to the evaluating and analysis result, discover the weak link of the system and give the improvement comment.

The Indexes of Systems Evaluating

Generally speaking, after the new system is completed, each kind of indexes should be evaluated comprehensively. But in the different stages of systems life cycle, the evaluating can be carried on partial indexes according to differently emphasis.

The systems evaluating indexes are the basis of systems evaluating and contrast analysis on new and old systems. For a management information system, the evaluating indexes may be divided into the economic indexes, performance indexes and management indexes.

Economic Indexes

- Systems expense, referring to sum of the development cost and the operating cost;

- Systems income, like the reduction of service expense, the enhancement of productivity, the drop of cost, the reduction of stock fund and the save of management charge and so on;
- Investment reclaim period;
- The scale and expense of systems reserve demand, the proportion of systems maintenance cost in total cost.

Performance Indexes

- Mean time of systems without failure;
- On-line response time, volume of goods handled or processing speed;
- Systems utilization factor;
- Operation convenience and flexibility;
- Security and secrecy;
- Data accuracy;
- The system extensibility.

Management Indexes

- The attitude of the leaders and the administrative personnel to the system;
- The attitude of the management information system user to the system;
- The evaluating of the external environment to the system.

Systems Evaluating Contents

Systems evaluating contents include:

- The realization situation of systems goals;
- Systems resources utilization factor;
- Systems security and secrecy;
- Systems usability;
- Systems maintainability;
- Systems cost;
- Completeness of systems management.

After the management information system is put into use, the application environment will change and management science and information technology level will be enhanced with the deep application of MIS. It is necessary to evaluate MIS unceasingly. We can have an explicit understanding to the current condition of the system. On the other hand, it also will prepare for the system future development and enhancement.

Notes

1. Moreover, the test goal should be to discover each kind of latent mistakes and flaws of the system by the least time and manpower.

句意：此外，系统测试的目的应该是以最少的时间和人力找出系统的每一种潜在的错误

和缺陷。

2. Probably the most common problem which arises during system integration is interface problems where sub-systems and components do not interact as anticipated by their developers.

句意：系统集成中可能出现的最常见的问题就是接口问题，即正如开发者所预想的那样，子系统或者单元不能相互作用。

3. The so-called systems evaluating is to estimate, inspect, test, analyze and examine the performance of management information systems.

句意：所谓的系统评价就是对管理信息系统的性能进行估计、检查、测试、分析和审查。

Exercises

1. What does system testing include?
2. What are the targets of system evaluating? And how to achieve these targets?
3. How many indexes are used in system evaluating?
4. Give a brief introduction to enterprise resources planning.
5. What are the advantages and disadvantages of enterprise resource planning?

Chapter 10 Human Resources Management

Unit 1 Organizational Design

Aim of HRM

Human Resource Management (HRM)[1] aims at recruiting capable, flexible and loyal people, managing and rewarding their performance and developing their key qualifications. Many people find HRM to be a vague and elusive concept-not least because it seems to have a variety of meanings. This confusion reflects the different interpretations found in articles and books about human resource management. HRM is an elastic term, it covers a range of applications that vary from book to book and organization to organization.

A HRM expert has distinguished between hard and soft forms of HRM, typified by the Michigan and Harvard models respectively. Hard HRM focuses on the resource side of human resources. It emphasizes costs in the form of Headcounts and places control firmly in the hands of management. Their role is to manage numbers effectively, keeping the workforce closely matched with requirements in terms of both bodies and behavior. Soft HRM, on the other hand, stresses the human aspects of HRM. It concerns communication and motivation. People are led rather than managed. They are involved in determining and realizing strategic objectives.

Necessity of Organizational Design

The HRM system includes a unified job organization structure and a banded salary structure. The job organization and pay band salary structure incorporates the wide variety of work performed.

The method in which jobs can be organized changes from a classification system consisting of over 1, 000 job classes to a unified one of approximately 300 broad job Roles. The job organization structure consists of 7 Occupational Families, approximately 60 Career Groups and approximately 300 Roles. Each Role describes the broad array of similar positions that are reflective of different levels of work

within a Career Group or occupational field and are assigned to one of the pay bands. The approximate 300 Roles cover the array of jobs of the classified workforce. All agency unique and central agency specific job classifications have been consolidated into this job organization structure. [2]

Job Organization Structure

The job organization structure is generally arranged into Occupational Families, Career Groups and Roles.

Occupational Families

An Occupational Family is a broad grouping that includes jobs that share similar vocational characteristics. The primary criterion for designation to a particular Occupational Family relates to the nature and type of work performed. The 7 Occupational Families include: Administrative Services, Engineering and Technology, Natural Resources and Applied Sciences, Health and Human Services, Educational and Media Services, Trades and Operations, and Public Safety.

Career Groups

A Career Group is a major subgroup of the Occupational Family that identifies a specific occupational field common to the labor market (e. g. Equipment Repair, Financial Services, Information Technology, Architecture and so on).

Roles

A Role describes an array of similar positions that are a reflection of different levels of work or career progression within a Career Group. Roles are intended to be very broad with a single Role including several former job classifications. For example, the Accountant (grade 9), Budget Analyst (grade 10), Auditor-Internal (grade 11) and Auditor-External (grade 11) in the former classification system are consolidated into one Role. Each Role is assigned to a specific pay band within the salary structure. The number of Roles in a Career Group varies from one Career Group to another. Additionally, some Career Groups have dual tracks for career advancement. For example, a Career Group may have a non-management track with entry, senior, and expert Roles and a management track with manager and director Roles. The manager's Role may be parallel to and within the same pay band as the staff expert Role.

Career Group Description

A Career Group Description specifies the nature and type of work associated with a particular occupational field and identifies the series of Roles within a Career Group. The Department of HRM maintains the master file of Career Group Descriptions.

Position

A position defines the specific core responsibilities, duties and any special assignments assigned to an employee. Through an analysis of the core responsibilities and special assignments (job evaluation process), a position is assigned to a Role. Each position is assigned the following titles:

Role Title is the formal state title that the employee's position has been assigned. The Role Title should be used for state reporting purposes.

Salary Reference Title is a descriptive title commonly understood and widely recognized in the labor market. The Salary Reference Title will be used for market surveying purposes and may be linked to the Standard Occupational Classification (SOC)[3] System.

Work Title is an agency-specific, or functional title, that is descriptive of the overall purpose of a position. Agencies may use Work Titles in conjunction with the employee's formal Role Title to help improve the recruitment process. Work Titles and Salary Reference Titles may be the same.

Employee Work Profile

The Employee Work Profile (EWP)[4] is the official state form that lists the core responsibilities, duties and any special assignments assigned to a specific position and incorporates the employee performance plan. Agencies may continue to use the Position Description and Performance Evaluation forms or develop their own forms as long as they include all the data elements and information contained in the EWP. The EWP is the principle source document for evaluating and allocating the position to the appropriate Role.

Instructions for Completing EWP

The EWP is a combination of the employee work description, performance plan, and evaluation assessment. Generally speaking, it includes:[49]

Position Identification Information

(1) Position Number: Enter assigned position number.

(2) Agency Name & Agency Code, Division/Department: Enter agency name and agency code; division or department name as appropriate.

(3) Work Location Code: Enter the position's location code.

(4) Occupational Family & Career Group: Enter the assigned occupational family and career group.

(5) Role Title & Code: Enter the position's Role title and code.

(6) Pay Band: Enter the pay band to which this role is assigned.

(7) Work Title: Enter the employee's work title if used (optional field).

(8) SOC Title & Code: Enter the assigned SOC title and code.

(9) Level Indicator & Employees Supervised: Check the appropriate box for employee, supervisor, or manager. Additionally, indicate if employee supervises two or more Full-Time Equivalent (FTE)[5] employees.

(10) FLSA Status: Check the appropriate box to designate the position as exempt or non-exempt under the *Fair Labor Standards Act* (FLSA)[6].

(11) Supervisor's Position Number: Enter the supervisor's assigned position number.

(12) Supervisor's Role Title & Code: Enter the supervisor's role title and code.

(13) EEO Code: Enter the appropriate Equal Employment Opportunity (EEO)[7] code.

(14) Effective Date: Enter the date the Employee Work Profile is effective (normally the date the position is established or changes are made to the work assignments).

Work Description & Performance Plan

(1) Organizational Objective: A brief statement describing how the position links to the work

unit, division or agency's objectives. This statement helps the supervisor and employee align the position's work assignments and priorities to agency-desired outcomes and results.

(2) Purpose of Position: A brief description of the reason the position exists. This statement should link to the organizational objective and capture the most important service or product expected from the employee in the position. This statement gives the reader a good idea of the purpose of the position without going into detail.

(3) Capabilities: A description of the expertise required to successfully perform the work assigned to the employee. It may be used in hiring new employees or to describe the competency or skill level of the responsibility.

(4) Core Responsibilities (A-F): Core responsibilities are defined as primary and essential to the work performed and are written as broad sets of major duties or functions. The core responsibilities must provide sufficient information to assign the position to the proper Role, determine FLSA exemption status, and provide a basis for performance evaluation.

(5) Measures for Core Responsibilities (A-F): Identify the qualitative and/or quantitative measures against which each responsibility will be evaluated.

(6) Special Assignments (G-H): Special assignments are considered brief in nature and typically are not extended beyond the performance period. Statements should be brief and do not have to include every detail of the assignment.

(7) Measures for Special Assignments (G-H): Identify the qualitative and/or quantitative measures against which each assignment will be assessed.

(8) Agency/Departmental Objectives (I-L): Objectives are defined as strategic business objectives to achieve goals set by the agency or division/department. They also may include behavioral competencies that are critical to the employee's success.

(9) Measures for Agency/Departmental Objectives (I-L): Identify the qualitative and/or quantitative measures against which each objective will be assessed.

Employee Development Plan

(1) Personal Learning Goals: List any learning goals identified by the employee and/or the supervisor.

(2) Learning Steps/Resource Needs: Indicate specific steps that need to be taken and by whom to accomplish the learning goals. This may include training, coaching, or other learning methods.

Review of Work Description/Performance Plan

The Comments, Signature, Print Name & Date of Employee, Supervisor and Reviewer.

Employee/Position Identification Information

(1) Position Number: Enter assigned position number.

(2) Agency Name & Agency Code; Division/Department: Enter agency name and agency code; division and department name as appropriate.

(3) Employee Name: Enter employee's full name.

(4) Employee ID Number: Enter unique ID number to identify employee.

Performance Evaluation (agencies may define the components)

(1) Core Responsibilities-Rating Earned (A-F): Check the appropriate rating earned by the employee during the performance cycle.

(2) Core Responsibilities-Comments on Results Achieved (A-F): Describe the employee's performance including documentation to support the earned rating.

(3) Special Assignments-Rating Earned (G-H): Check the appropriate rating earned by the employee.

(4) Special Assignments-Comments on Results Achieved (G-H): Describe the employee's performance including documentation to support the earned rating.

(5) Agency/Department Objectives-Rating Earned (I-L): Check the appropriate rating earned by the employee (optional).

(6) Agency/Department Objectives-Comments on Results Achieved (I-L): Describe the employee's performance including documentation to support the earned rating.

(7) Other Significant Results for the Performance Cycle: Record any significant aspects of the employee's job performance that are not addressed elsewhere in the evaluation. Only include comments that are related to the employee's job performance.

Employee Development Results (agencies may define the components)

Year-End Learning Accomplishments: Summarize accomplishments related to the personal learning goals that were set at the beginning of the cycle.

Overall Results Assessment and Rating Earned

Overall Rating Earned: Check the appropriate overall rating earned by the employee during the performance cycle.

Review of Performance Evaluation

The Comments, Signature, Print Name & Date of Employee, Supervisor and Reviewer.

Confidentiality Statement

Allows an agency to identify confidentiality as a critical organizational value and to establish clear consequences if confidentiality is violated. This section should be used only in circumstances where information is protected and does not apply to information that is released according to agency procedures. Annual requirements provide a method to easily gather and track certain agency-specific information on an annual basis.

Physical/Cognitive Requirements

Documents essential and marginal job functions of the position for use when responding to requests for modification or accommodation. This information should be maintained as part of the position information. Any medical information used to make job modifications or accommodations must be maintained separately from the employees' personnel files.

Guide for Designing an EWP

The EWP should be designed to have a page-break just before Review of Work Description/Performance Plan. This allows Position Identification Information and Work Description & Performance

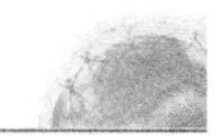

Plan to be easily copied and shared with other agencies and Department of HRM, without sharing personal development planning or performance evaluation information. This page-break might also improve agency file organization. It is recognized that some agencies may want to adopt other formats and to keep the Position Description and Performance Management forms totally separate. Whatever format is used, the following elements are required:

Position Identification Information

Position identification information includes: Position Number; Agency Code & Name, Division/Department Name; Work Location Code; Occupational Family & Career Group; Role Title & Code; Pay Band; Work Title (optional); SOC Title and Code; Level Indicator & Employees Supervised; FLSA Status (for exempt/partially exempt positions, include Exemption Test used, if not retained elsewhere); Supervisor's Position Number; Supervisor's Role Title & Code; EEO Code; Effective Date.

A Work Description & Performance Plan

A Work Description & Performance Plan is required, however the structure and elements as listed in the prototype are not required. Information in this section must be sufficient to assign the position to the proper Role, determine FLSA status, and provide a basis for performance evaluation.

An Employee Development Plan

An Employee Development Plan is required, however the structure and elements as listed in the prototype are not required.

Review of Work Description/Performance Plan

Signatures designating Review of Work Description/Performance Plan are required of employee, supervisor, and reviewer. The signature section should be on a separate page so as to facilitate filing and protection of employee privacy should the position description be shared.

Employee/Position Identification Information

The following information is required as part of the Performance Evaluation sections; Position Number; Agency Name & Code; Division/Department; Employee Name; Employee ID Number.

Performance Evaluation

A Performance Evaluation is required, however the structure and elements as listed in the prototype are not required. This is a confidential section.

Employee Development Results

A section addressing Employee Development Results is required, however the structure and elements as listed in the prototype is not required. This is a confidential section.

Overall Results Assessment and Rating Earned

Overall Results Assessment and Rating Earned are required as listed in the prototype. This is a confidential section.

Review of Performance Evaluation

Signatures designating Review of Performance Evaluation are required of employee, supervisor, and reviewer. All of the elements listed are required to notice the Improvement Needed/Substandard Performance, however the format can differ by agency.

Notes

1. Human Resource Management (HRM) 人力资源管理

2. All agency unique and central agency specific job classifications have been consolidated into this job organization structure.

句意：所有专门代理机构和中枢代理机构的特殊工作的分类已经被统一到这个工作组织结构中。

3. Standard Occupational Classification (SOC) 标准职业分类
4. Employee Work Profile (EWP) 职工工作概况
5. Full-Time Equivalent (FTE) 相当于全时工作的，相当于全职的
6. Fair Labor Standards Act (FLSA) (美国) 公平劳动标准法
7. Equal Employment Opportunity (EEO) 均等就业机会

Exercises

1. What is the aim or function of a HRM system?
2. What does a job organization structure consist of ?
3. What does a completing EWP consist of ?
4. How to design an EWP?

Unit 2 Performance Measurement

Aim of Performance Measurement

The performance management program used in the Compensation Management System retains selected features of the original Employee Incentive Pay Program (EIPP)[1] but expands the concept of linking employee performance to pay. The performance management program has been designed to insure that increases approved by the Governor and the General Assembly can be appropriately distributed based on the employee's performance rating. The pay band structure allows for performance increases without having to change the band structure.

The expanded features of the performance management system include fewer performance rating levels, an extended probationary period for new employees, employee self-assessment and employee feedback on supervisor's performance. The performance program is flexible and allows agencies to design optional features to effectively meet their agencies' needs.

Basic Terms of Performance Management

Simply put, performance management includes activities to ensure that goals are consistently being met in an effective and efficient manner. Performance management can focus on performance of an organization, a department, processes to build a product or service, employees, etc. Information in this

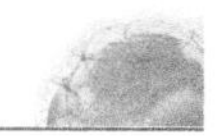

topic will give some sense of the overall activities involved in performance management. The basic terms include the following:

Domain

The domain is the focus of the performance management effort, e. g. , the entire organization, a process, subsystem or an employee. A subsystem could be, e. g. , departments, programs (implementing new policies and procedures to ensure a safe workplace; or, for a nonprofit, ongoing delivery of services to a community), projects (automating the listing process, moving to a new building, etc.), or teams or groups organized to accomplish a result for an internal or external customer. A process produces a product or service for internal or external customers, and usually cuts across multiple subsystems. Examples of processes are market research to identify customer needs, product design, product development, budget development, customer service, financial planning and management, program development, etc. The final domain is that of employee performance management. The term domain is not widespread across performance management literature.

Results

These are usually the final and specific outputs desired from the domain. Results are often expressed as products or services for an internal or external customer, but not always. They may be in terms of financial accomplishments, impact on a community, etc. Results are expressed in terms of cost, quality, quantity or time.

Measures

Measures provide specific information used to evaluate the extent of accomplishment of results. Measurements are typically expressed in terms of time, quantity, quality or cost. Results are a form of measure.

Indicators

Indicators are also measures. They indicate progress (or lack of) toward a result. For example, some indicators of an employee's progress toward achieving preferred results might be some measure of an employee's learning (usually expressed in terms of areas of knowledge or specific skills) and productivity (usually measured in terms of some number of outputs per time interval).

Organization's Preferred Goals

These are usually overall accomplishments desired by an organization and are often established during strategic planning. The level of specificity of goals depends on the nature and needs of the organizations. Typically, the more specific the goals, the clearer the understanding of goals by the members of the organization.

Organization's Preferred Results

The performance management process often includes translating organizational goals to be in terms of results, which themselves are described in terms of quantity, quality, timeliness or cost.

Aligning Results

Performance management puts strong focus on ensuring that all parts of the organization are working as efficiently and effectively as possible toward achieving organizational results. Therefore, the results of all parts of the organization should be aligned with the overall preferred results of the organization. Aligning results often includes answering questions such as "Does the domain's preferred results contribute to achieving the organization's preferred results? How? Is there anything else that the domain could be doing to contribute more directly to the organization's goals?"

Weighting Results

Weighting results refers to prioritizing the domain's preferred results, often expressed in terms of a ranking (such as 1, 2, 3, etc.), percentage-time-spent, etc.

Standards

These specify how well a preferred result should be achieved by the domain. For example, "meets expectations" or "exceeds expectations".

Performance Plan

The plan usually includes at least the domain's preferred results, how the results tie back to the organization's preferred results, weighting of results, how results are measured and what standards are used to evaluate results.

Ongoing Observation, Measurements and Feedback

These activities include observing the domain's activities in terms of progress toward preferred results, comparing progress to the preferred performance standards and then providing ongoing feedback (useful, understandable and timely information to improve performance) to the domain.

Performance Appraisal (or Review)

In its most basic form, performance appraisal (or review) activities include documenting achieved results (hopefully, by also including use of examples to clarify documentation) and indicating if standards were met or not. The appraisal usually includes some form of a plan supposed to improve insufficient performance.

Rewards

The performance review process usually adds information about rewarding the employee (s) if performance met or exceeded standards. Rewards can take many forms, e.g., merit increases, promotions, certificates of appreciation, letters of commendation, etc.

Performance Gap

This represents the difference in actual performance shown as compared to the desired standard of performance. In employee performance management efforts, this performance gap is often described in terms of needed knowledge and skills which become training and development goals for the employee.

Performance Development Plan

Typically, this plan conveys how the conclusion was made that there was inadequate performance, what actions are to be taken and by whom and when, when performance will be reviewed again and how. Note that a development plan for employee performance management may be started for various reasons other than poor performance.

Principles of Performance Management

The performance management was developed based on the following principles:

- To identify individual and/or team objectives and measures linked to the agency's mission and strategic objectives;

 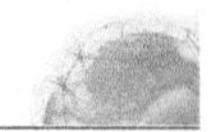

- To promote employee and career development through creating an environment of learning and quality improvement through training, coaching, counseling and advising;
- To provide open and honest periodic evaluations of employees' performance;
- To administer financial rewards based on distinctions in performance with the highest level of award directed to "Extraordinary Contributor" and a lesser amount directed to "Contributor".

The Performance Management Program has intentionally been designed to allow agencies a great deal of flexibility in planning their performance management systems to meet their unique organizational needs. Therefore, agency management will need to make a number of decisions prior to implementing their Performance Management Program. The agency will need to describe the design and the implementation strategy of their Performance Management Program in their Agency Salary Administration Plan.

Ratings of Performance Management

It is important to emphasize that evaluation of employees' performance should be done on a continuous basis by providing verbal and/or written feedback throughout the performance cycle. The Commonwealth's Performance Management Program has 3 performance ratings that are used to rate performance on individual measures; and are the composite rating for an employee's overall performance. These ratings are described below.

Extraordinary Contributor

This rating recognizes work that is characterized by exemplary accomplishments throughout the performance cycle and performance that considerably and consistently surpasses the criteria of the job function. To be eligible to receive an overall rating of Extraordinary Contributor, an employee must receive at least one documented Acknowledgment of Extraordinary Contribution form during the rating cycle. However, the receipt of this form does not guarantee or necessarily warrant an overall annual rating of Extraordinary Contributor.

Contributor Rating

This rating recognizes work that is at or above the performance standards by achieving the criteria of the job function throughout the performance cycle. Employees at this level are achieving the core responsibilities and performance measures as outlined by the supervisor.

Below Contributor

This rating recognizes job performance that fails to meet the criteria of the job function. An employee who receives at least one Need Improvement/Substandard Performance form may receive an overall rating of Below Contributor on the annual rating. An employee cannot be rated Below Contributor on the annual evaluation if he or she has not received at least one Need Improvement/Substandard Performance form during the performance cycle.

Process of Performance Management

1. Development of Employee Work Profile/Performance Plan

The general Performance Management Program is designed to be flexible and allow agencies to design optional features that address their specific organizational needs more appropriately. It is recommended that agencies use a combination form similar to the sample Employee Work Profile. However, agencies may design and use form (s) that meet their unique needs as long as each employee has a work description with the required position identification information and a performance plan. The work description must be sufficient enough to assign a position to the proper Role, determine the FLSA status and provide the basis for the performance evaluation.

2. Performance Planning Meeting

Generally, a supervisor should schedule a Performance Planning Meeting with the employee within 30 days from the beginning of each performance cycle. At this meeting, the employee and supervisor should review the core responsibilities and special assignments from the previous cycle and update them for the new cycle. If the employee is new, the position changes or has significant changes to the core responsibilities, the supervisor and employee should meet to discuss these changes within 30 days of the employee's begin date or date of the job changes.

3. Information Gathering for the Performance Evaluation

In order to effectively evaluate performances and reflect the actual performance of the employee, it is necessary to collect and document information on a continuous basis. Supervisors should use a variety of sources when gathering evaluation information. Sources may include direct observation of employee behaviors and work products by the supervisor and information requested from peers, customers, subordinates, and other supervisors who interact and work with the employee. The information is most relevant and valuable when these additional sources possess a clear understanding of the employee's core responsibilities. Employees should be informed that potential sources may be used in the evaluation process at the beginning of the performance cycle.

4. Feedback During the Performance Evaluation Cycle

Supervisors should advise, coach and reinforce progress toward expected results and outcomes and address areas of concern as they arise. Effective management of performance involves providing continuous formal (written documentation) and/or informal (verbal) feedback to employees throughout the entire performance cycle. Formal feedback may be accomplished through Interim Performance Evaluations or written notification for exemplary accomplishments or substandard performance.

5. Completing the Performance Evaluation

The supervisor must have performance evaluations finalized for employees that have completed a full 12-month performance cycle (e. g. October 25th through the next October 24th) by October 24th, but not before August 10th. In completing the performance evaluation and arriving at an overall performance rating, the supervisor should take into consideration how successful the employee was in meeting the criteria established by the performance measures and the length of time the employee performed in their job. If an employee was absent from work for a significant portion of the performance year, the percentage increase may be impacted. Employee absences due to compensatory time, overtime leave, workers' compensation, military, Family and Medical Leave and Short Term Disability should not influence the employee's overall performance rating. Reviewer Approval of the Performance Evaluation.

6. Appeals

When an employee disagrees with the evaluation and cannot resolve the disagreement with his su-

pervisor, the employee may appeal to the reviewer for reconsideration. The employee must make this appeal in writing within 10 workdays of the initial evaluation meeting. The reviewer should discuss the appeal with the supervisor and the employee. After discussion of the appeal, the reviewer should provide the employee with a written response within 5 workdays of receipt. The response should indicate the reviewer's conclusion of the performance evaluation. In addition, agencies may develop their own appeals process for the reconsideration of employee performance evaluations. The appeals process should be outlined in the Agency Salary Administration Plan.

7. Performance Evaluation Formula

The General Assembly and the Governor annually determine the statewide average performance increase for the Commonwealth's workforce. Agencies may not supplement the funding provided by the General Assembly and Governor for employee performance increases. The agency may elect to distribute the performance increase based on the agency as a whole or proportion the increase based on designated sub-agencies. A sub-agency is a designation an agency may use to sub-divide into smaller organizational units to fairly and equitable distribute performance increases based on available funding. This designation would normally occur at a level where the responsible manager exercises full authority over both personnel and budget management. The actual increase may vary from one agency or sub-agency to another based on the formula used within these limits and the number of employees in the agency or sub-agency that are rated as contributor. All agencies or sub-agencies (within an agency) do not have to give the same percentage increase to employees rated as Contributors or Extraordinary Contributors. However, employees that receive the same rating within the agency or sub-agency must receive the same performance increase amount.

Key Benefits of Performance Management

Generally, the key benefits of performance management include the following aspects:

1. Performance Management Focuses on Results, Rather Than Behaviors and Activities

A common misconception among supervisors is that behaviors and activities are the same as results. Thus, an employee may appear extremely busy, but it is not contributing at all toward the goals of the organization. An example is the employee who manually reviews completion of every form and procedure, instead of supporting automation of the review. The supervisor may conclude the employee is very committed to the organization and works very hard, thus, deserving a very high performance rating.

2. Aligns Organizational Activities and Processes to the Goals of the Organization

Performance management identifies organizational goals, results needed to achieve those goals, measures of effectiveness or efficiency (outcomes) toward the goals, and means (drivers) to achieve the goals. This chain of measurements is examined to ensure alignment with overall results of the organization.

3. Cultivates a System-wide, Long-term View of the Organization.

An effective performance improvement process must follow a systems-based approach while looking at outcomes and drivers. Otherwise, the effort produces a flawed picture. For example, laying off people will likely produce short-term profits. However, the organization may eventually experience reduced productivity, resulting in long-term profit loss.

4. Produces Meaningful Measurements

These measurements have a wide variety of useful applications. They are useful in benchmarking, or setting standards for comparison with best practices in other organizations. They provide consistent basis for comparison during internal change efforts. They indicate results during improvement efforts, such as employee training, management development, quality programs, etc. They help ensure equitable and fair treatment to employees based on performance.

Notes

1. Employee Incentive Pay Program (EIPP) 职工激励（奖励）工资计划

Exercises

1. What is the main content of Performance Management?
2. What is the principle of Performance Management?
3. How to rate a Performance Management Program?
4. How to describe the process of Performance Management?
5. What are the benefits of Performance Management?

Chapter 11
Science-Technology Literature Search

Unit 1 Introduction to Literature

Definition of Literature

A literature is defined as a carrier that records knowledge, which includes the body of knowledge, symbol of information and the material carrier.[1] There are four properties of the literature as following:

(1) Knowledge and information. It is the basic function for a literature to convey information and record knowledge.

(2) Objectivity. Information and content represented in a literature must resort to some symbol of information as well as some material carrier, by which the literature can be preserved and transported for a long time.[2]

(3) Manual recording. The information represented in a literature is recorded manually rather than adhered inartificially.

(4) Development. With the progress of the ability for mankind to record information, there are huge volumes of literatures and their categories are diversified. At the same time, some of the literatures have been out of date and the whole life is reduced, which causes regular movements of literatures.

Kinds of Literature

There are many ways to classify literatures. In this section, we mainly describe the kinds of literature which refers to Eileen Pritchard and Yu jing Fan. The literature is divided into two classes: primary sources and secondary sources.

1. Primary Sources

(1) The scientific journal. Modern primary journals contain articles, which are reports research or original observations. The journal article may have evolved from the paper prepared for oral delivery at meetings of scholarly societies, or perhaps from prize essays which early societies developed to stimulate scientific research.

The modern scientific journal article often has a set of format. There may or may not be a summary, or abstract, at the beginning of article. There is usually an introduction which states the problem examined in the article. The discussion brings the results together, evaluates them, and interprets them in the light of other research. There is usually a reference section at the end of the article listing the literatures cited in work.

(2) Trade journals. Practical information related to persons in industry is conveyed in the trade journal. The content includes business news, product information, advertising, and trade articles. The journals can provide a great deal of information on current trends in technology, and are useful to persons seeking orientation to a vocation.

(3) The technical report. Technical reports are accounts of work done on research projects; they are written to provide information to employees and other research works. A report may emanate from completed research or on-going research projects. Private companies and associations use reports internally for communication within the organization, and occasionally for communication within the organization, support many technical reports by means of grants and government contracts.

(4) Proceedings. Scientists present original research findings and review articles at professional meetings. Often these are published and distributed in various forms. The meetings may be referred to as symposia, conferences, institutes, workshops, or colloquia. They provide an important channel of communication for scientists and an important source of information for researchers.

(5) Abstracts of research in progress. Abstracts may be primary sources when they are used to report research in progress presented at a meeting before the journal research article appears.

(6) Dissertations. Doctoral dissertations are another primary source of scientific publication. In the science, the awarding of a Ph. D. degree usually requires completion of a major monograph including extensive experimentation, reporting on the results, and suggesting future implications. The dissertations are kept in libraries at the home schools of the doctoral candidates, and we can get the full text from the network digital resource in most of university libraries web. [3]

(7) Patents. Patents are rights granted by law for protection of inventions or discoveries. Patent specifications describe designs, methods and processes of the invention, and are, therefore, an important source of information for the engineer, physicist, chemist, and other researchers. To locate patents there are commercial indexing services such as Chemical Abstracts that include patents, and government patent indexes. Recently, database searching has become useful in examination of patent literature.

(8) Standards. Standards are requirements for the quality or size or shapes of industrial products. They also comprise recommendations for methods and processes in manufacturing. Standards are prepared by a variety of trade association, national and international bodies. Some types of standards include quality and measure recommendations, testing materials, and definitions of trade terms.

2. Secondary Sources

(1) Abstracts. There are many services which abstract and index technical publications so that researchers can select important papers quickly in their field of interest.

(2) Reviews. Review articles distill the existing knowledge relevant to a particular subject into a compact, accessible form. A typical review article focuses on important advances which have been made in a specialty, evaluate research, indicates where gaps in knowledge exist, and provides a com-

prehensive bibliography on the subject.

(3) Specialized books. There are many important books that contain authoritative information and are considered to be basic in the field.

(4) Reference books. In many cases, specific factors or a summary of a topic are all that is required. Handbooks, manuals, encyclopedias, and dictionaries perform this function.

The Reasons to Do a Literature Search

For the scientists and researchers, the problem of keeping up with the growing volumes of new knowledge threatens both productivity and research quality. For a student, your literature review forms the central core upon which you hang the argument of your thesis. It is intended to present and synthesize the background to your research, to demonstrate that you have read appropriately within your subject area, and to show where your work fits into existing knowledge. For the paper published, it will form the first chapter of your paper, and without it, what you write will not fulfill the requirements for a paper. You will also need to be able to identify papers for assignments and project work. Therefore, the following are some main aims to do a literature search.

- Increases your own knowledge of the subject area;
- Helps you identify work already done or in progress that is relevant to your own;
- Prevents you from duplicating work already done (and helps to avoid accusations of plagiarism);
- Helps you avoid errors of previous research;
- Helps you choose or design your own methodology;
- Enables you to find 'gaps' in existing research and find a unique area for your own.

Notes

1. A literature is defined as a carrier that records knowledge, which includes the body of knowledge, symbol of information and the material carrier.

句意：文献是记录知识的载体，它包括知识部分、信息符号和物质载体。

2. Information and content represented in a literature must resort to some symbol of information as well as some material carrier, by which the literature can be preserved and transported for a long time.

句意：文献中所表述的信息和内容必须借助于某种信息符号和物质载体来进行长期的保存和传输。

3. Doctoral dissertations are another primary source of scientific publication. In the science, the awarding of a Ph. D. degree usually requires completion of a major monograph including extensive experimentation, reporting on the results, and suggesting future implications. The dissertations are kept in libraries at the home schools of the doctoral candidates, and we can get the full text from the network digital resource in most of university libraries web.

句意：博士论文是科技出版物另外一种主要来源。在科学上，博士学位的授予通常需要一个包括大量试验、结果报告和未来方向的专论。博士论文一般保存在博士生母校的图书馆中，我们可以通过大学图书馆网页来获得网络电子资源的博士论文全文。

Exercises

1. What is the meaning of literature?
2. How many kinds of literature?
3. Why to do a literature search?

Unit 2 Methods of Doing a Literature Search

Essence of Literature Search/Retrieval

A literature search is a systematic and thorough search of all types of published literatures in order to identify as many items as possible that are relevant to a particular topic.[1] In essence, literature retrieval can be defined as the searching for information in some databases.

The Tools to Do a Literature Search

The tools of literature search/retrieval are made up of description, bibliographic, index and appendix.

1. Description. It is an instructing part for user to use the retrieval tools, which includes the aim to organize the tools, scopes, date and retrieval methodologies, etc. A user should read this description before doing a search.

2. Bibliographic. It is made up of the name, author and so on, which is main body of retrieval tools. Each literature has an item of bibliographic, which makes up the literature warehouses. Literature retrieval's aim is to search the relevant records in literature warehouse and gets some information from the original text and clues of searched literature. Therefore, the bibliographic is the main body of retrieval tools.

3. Indexes. In order to improve the efficiency of retrieval, various indexes are used, such as subject index, classification index, author index, company/org index, etc. Although it is an assistant way to retrieve literatures, indexes play a great role in practice.[2]

4. Appendix. Appendix is supplementary to retrieval tools, which includes the categories of journals, acronyms, interpretation, terminology, etc.

At present, the tools of retrieval are classified catalogue, title, abstract and index.

(1) Catalogue. Catalogue is composed by items on the base of arraying rules, which mainly includes:

- The name of the book or the journal;
- Author;

- Item of publication;
- Accessory;
- Abstract.

(2) Subject. Subject mainly describes the external characters of a literature, such as the title, author, publication, date, the kind of a literature and language, etc.

(3) Abstract. According to the aims and the degree of detail, abstract can be mainly classified three kinds: informative abstract, indicative abstract and informative-indicative abstract.

(4) Index. There are four kinds of index, the first is subject index, the second is classification index, the third is author index, and the fourth is number/code index.

- Subject index: The subject can represent the main idea of a literature in essential;
- Classification index: the classification is that the number of classification is arrayed according to the body of a literature, and the names of classification are given in according to the number of classification;
- Author index: this index is arrayed by the names of literature's authors, and the abstract number is provided following the authors;
- Number/code index: it is mainly used in science-technology reports, patent literatures, standard literature, etc.

Formulating a Basic Search Strategy

When implementing a research project, one must carefully analyze a question appropriate to the topic and methods to solve the question. The process of doing a basic search is shown in the Figure 11-1.

1. Definition of terms related to a topic

An understanding of the chosen subject is prerequisite to searching in bibliographic sources. Dictionaries containing scientific and technical terms can clarify basic terminology.

2. Preliminary background information

To gain general background information and to understand how a subject relates to the various fields of knowledge, an encyclopedia and/or textbook can provide a good introduction to the subject.

3. Materials from basic indexes

Journal articles provide most of the original information. Indexes and abstracts guide the researcher to journal articles; sometimes review articles, and symposia. Indexes may also include listing of books, reports, or even dissertations.

4. Materials from specialized indexes—current information

Current awareness journals and indexes are important because they cover materials too recent to be indexed in the more comprehensive abstracts and indexes. A good paper or thesis should include a search for the most current materials.

5. Materials from specialized indexes—reviews and meetings

Since the review and symposia papers are not as vital as original research in journals, their indexes are placed in step 5 of the first search strategy diagram. A more experienced searcher may wish to search for reviews and symposia first.

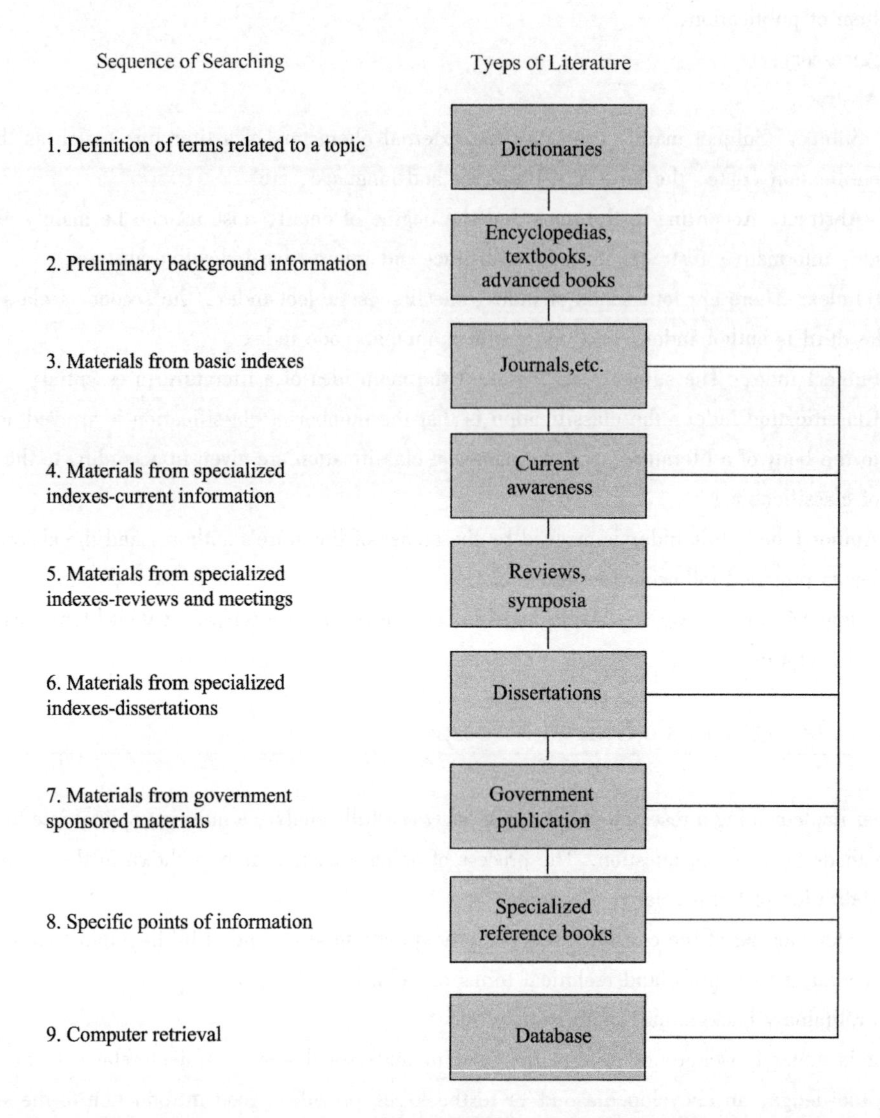

Figure 11-1 The Process of Doing a Literature Search

6. Materials from specialized indexes—dissertations

It is probably not necessary to look for dissertations when writing a term paper. Dissertation can be useful if the topic needs comprehensive coverage. If the material is important, it will usually appear in journal literature.

7. Information from government materials

The China government is an example of publishers of specialized governmental materials which provide information to someone writing a paper or doing research. Government publications are particularly important in the field of agriculture. Statistics are also found in government publications, since government agencies collect many statistics.

8. Specific points of information

The library user occasionally needs one fact or statistical information derived from a table. Many

reference books contain short explanations or collections of facts and data.

9. Computer retrieval—database

If the major part of the significant material is within the last ten years, the researcher may wish to utilize a database search. A database search may save much searching time in library.

Evaluation of the Performance

1. Characteristics of Good Measures

Measures are based on the performance criteria identified. A good measure is one that reflects the essence of the criterion and that implies a unit for quantification. Validity and reliability are the two most important qualities of measures.

Validity relates to the accuracy and representativeness of the data measured. It is the property linking what is actually measured to what the concept purports to measure. In other words, the measure should reflect the real meaning of the concept to be measured. [3]

Reliability is the degree of stability or consistency of the data collected over time. That is, if the same data collection procedure is repeated under the same conditions, would the same scores be obtained? It is the property of stability in recording the results of measuring. It is directly concerned with the accuracy and precision of the measuring instrument. [4]

2. Measure units

The two most well known performance measures are recall and precision. Recall measures the proportion of the relevant documents actually retrieved. That is, assuming that the collection contains 100 items on the subject being sought, how many of the 100 have been retrieved? Precision relates to the percentage of relevant documents contained in the retrieved set. In other words, when one retrieves 50 citations, how many of these are actually relevant? Naturally, it is assumed that relevance has been established by someone or some yardstick. No matter how limited these measures are, recall and precision have become the standards used by most evaluation experiments.

$$PECALL = \frac{\textit{number of relevant documents retrieved}}{\textit{total number of relevant documents in the file}}$$

$$PRECISION = \frac{\textit{number of relevant documents retrieved}}{\textit{total number of documents retrieved from the file}}$$

In probabilistic terms, recall is an estimation of the conditional probability that an item will be retrieved given that it is relevant to the query. Precision is an estimation of the conditional probability that an item will be relevant given that it is retrieved. [5]

To take account of the size of the collection from which retrieval is conducted, specificity was another measure proposed by a Western Reserve University research group in the 1960s. Specificity measures how well the systems are able to reject non-relevant documents with respect to the size of the collection.

Fallout has also been suggested as another parameter of a search. It is the probability of a false drop. It is another measure attempting to incorporate the effects of the file size. Unfortunately, both specificity and fallout are fairly insensitive measures when the retrieved set is quite small in comparison with large files.

$$Specificity = \frac{number\ of\ nonrelevant\ documents\ not\ retrieved}{total\ number\ of\ nonrelevant\ documents\ in\ the\ file}$$

$$Fallout = \frac{number\ of\ nonrelevant\ documents\ retrieved}{total\ number\ of\ nonrelevant\ documents\ in\ the\ file}$$

Finally, efficiency was proposed as a single measure of system effectiveness on retrieving relevant and only relevant documents.

$$Efficiency = Recall + Specificity - 1$$

Electronic Databases for a Literature Search

One of the major sources that come to mind when doing a literature search is electronic databases. Taking over from the printed index, there are many databases available. Some databases are available freely on the Internet, while others are available by subscription only and you should explore their availability at your library.[6] Some key databases include EBSCO Host Research Database, Elsevier Database, Science Online, Springer Link, Kluwer Online, Cnki Database and Vip Database.

When we do a literature search by database, we should master some of the following techniques, which includes Keyword, Boolean logic, Truncation/wildcard, Spelling and alternative terms, Focusing a search[6].

1. Keywords

It is important to have keywords in mind before starting your search. It is best to think laterally about the search topic and consider all the possible synonyms and related areas of your search topic. If there are no keyword appropriate to your topic, then it is best to use free text searching. This is where words are retrieved anywhere in the reference including the abstract. It may therefore mean that some irrelevant records are retrieved that may merely mention the term you are looking for. In order to capture as many records as possible it is best to search using keywords from the thesaurus as well as free text searching.

2. Boolean logic

Boolean logic is important for joining, widening or excluding terms when conducting a search:

(1) AND should be used to join concepts to make a search more specific, e. g. Economic AND Evolution.

(2) OR should be used to widen a search when you require any set of words to be present or to specify, e. g. economic evolution OR institute evolution.

(3) NOT should be used to exclude words, e. g. economic evolution NOT Institute evolution.

(4) NEAR can also be used to search for terms together, e. g. chartered NEAR society NEAR physiotherapy.

(5) Quotation marks can also be used to search for a phrase, e. g. "chartered society of physiotherapy".

3. Truncation/wildcard

The process known as truncation, wildcard or word stemming lets you search on part of a word to

retrieve information on similar words and this can simplify searching.[7] A range of symbols can be placed after the word and this varies among databases.

4. Spelling and alternative terms

It is always important to consider different spellings of words. Alternative terms should also be borne in mind and these can be searched for using Boolean commands. For example one might search for ‘physiotherapy’ or ‘physical therapy’ in order to retrieve all the possible records for this topic.

5. Focusing a search

Depending on the search topic, a simple search can often retrieve a vast amount of records. There are ways in which you can narrow down your search. For example it is possible to:

Specify the language of information retrieved:

- Limit the time scale of a search by narrowing it to specific years;
- Search for a particular publication type;
- Search for review articles only;
- Search for words in the title of an article only;
- Limit by age or gender.

Notes

1. A literature search is a systematic and thorough search of all types of published literature in order to identify as many items as possible that are relevant to a particular topic.

句意：文献检索是为了系统而彻底地搜索所有类型的出版文献以便获得尽可能多的与特定主题相关的条目。

2. In order to improve the efficiency of retrieval, various indexes are used, such as subject index, classification index, author index, company/org index, etc. Although it is an assistant way to retrieve literatures, indexes play a great role in practice.

句意：为了提高检索效率，可以使用不同的索引，如主题索引、分类索引、著者索引、出版社索引等。虽然索引是检索文献的辅助手段，但它在实践中却扮演着重要的角色。

3. Validity relates to the accuracy and representativeness of the data measured. It is the property linking what is actually measured to what the concept purports to measure. In other words, the measure should reflect the real meaning of the concept to be measured.

句意：有效性与被检索数据的精确性和代表性相关。它是连接实际检索到的数据和概念上应被检索到的数据两者之间的属性。换句话说，检索应该反映将要被检索的数据的真正含义。

4. Reliability is the degree of stability or consistency of the data collected over time. That is, if the same data collection procedure is repeated under the same conditions, would the same scores be obtained? It is the property of stability in recording the results of measuring. It is directly concerned with the accuracy and precision of the measuring instrument.

句意：可靠性是多次检索数据的稳定性和一致性的程度，也就是说，如果在相同条件下重复检索相同的数据，其结果是否相同？它是表示检索结果稳定性的属性，并直接与检索工具的精确性有关。

5. In probabilistic terms, recall is an estimation of the conditional probability that an item

will be retrieved given that it is relevant to the query. Precision is an estimation of the conditional probability that an item will be relevant given that it is retrieved.

句意：从概率方面来说，（计算机）信息检索（能力）是对与给定检索相关的项目能被检索到的条件概率的估计值。

6. Some databases are available freely on the Internet, while others are available by subscription only and you should explore their availability at your library.

句意：有些数据库在互联网上可以免费使用，而大部分数据库主要通过订阅（付费）使用，或者可以在图书馆网页上查找相关电子数据库使用。

7. The process known as wildcard or word stemming lets you search on part of a word to retrieve information on similar words and this can simplify searching.

句意：采用通配符或单词填充的过程可以使你通过对单词的部分查询来检索与单词相关的信息，这个过程可以简化查询。

Exercises

1. What is the essence of literature search?
2. What are the characteristics of good measures?
3. What are the two most well known performance measures? Please give the specific functions.
4. Please give an example to do a basic literature search according to the above strategy.
5. What do key electronic databases used to retrieve science papers include?
6. What are the steps to do a literature search by electronic database?
7. Please try to use EBSCO, Springer Link, Cnki Databases to retrieve some information related to your paper.

Chapter 12
Scientific Literature Translation

Architecture Features of Scientific Literature

The architecture of scientific literature advocates preciseness, accurate concepts, strong logic, concise style of writing, outstanding emphases, neat and less changeable sentence type.[1] In a word, the architecture features of scientific literature are clear, accurate, concise and tight.

Grammar Features of Scientific Literature

Scientific literature is a carrier of science and technology, which is used to present the nature of science and technology objectively. It requires rigor logic and it should specify the viewpoints of authors objectively. Thus the present tense, indicative mood, modal verbs such as *can* and *may*, the passive voice, post modifiers are often used in scientific literature.

The Use of Present Tense in Scientific Literature

Scientific literature uses more present tense and the indicative mood of the verb to show general narration with timeliness, namely narrate the fact or the truth. Objectively to state definition, theorem, equation, formula and chart, etc. For example:

(1) The rotation of the earth on its own axis causes the change from day to night.

地球绕轴自转，引起昼夜的变化。

(2) AIDS is spread by direct infection of the bloodstream with body fluids that contain the AIDS virus.

艾滋病传播是由血液和带有艾滋病病毒的体液直接感染引起的。

(3) Force is a vector which has both magnitude and direction.

力是一种向量，它既有大小又有方向。

(4) Hooke law states that the strain developed is directly proportional to the stress producing it.

胡克定律表明，应变与引起应变的应力成正比。

The Use of Modal Verb

In scientific literature, the using frequency of modal verbs *can* and *may* is a little higher compared with others. This is because these two modal verbs can be used for expressing objective possibility, while others stress more on subjectivity.

(1) Anyone with a personal computer, a modem and the necessary software to link computers over telephone lines can sign on.

任何人只要有一台个人计算机，再有一个调制解调器和必要的软件，就可以把计算机连接到电话线上，然后就能申请上网。

(2) The modern digital computer is an electronic machine that can perform mathematical or logical calculations and data processing functions in accordance with a predetermined program of instructions.

数字计算机是能够完成数字和逻辑运算，并能够按照事先编制好的程序进行数据处理的电子设备。

The Use of Passive Voice

Passive voice is more widely used in scientific literature. The statistics shows, the passive voice of predicate in scientific literatures amounts to 1/3 at least. This is because scientific literature put more emphasis on stating the fact, method, performance and characteristic objectively. Compared with active voice, passive voice expresses more objectively, which helps to concentrate the readers' attention on narrating objects, reality or the course.

(1) Atoms can be thought of as miniature solar systems, with a nucleus at the center and electrons orbiting at specific distances from it.

原子可以被看成是一个微型的太阳系，原子核在其中心，而电子以一定的距离绕核做圆周运动。

(2) Attention must be paid to the working temperature of the machine.

应当注意机器的工作温度。

In addition, if the first person and the second person are used too much, it will cause the impression that the material is assumed subjectively. So try one's best to use the third person to narrate, adopt the passive voice.

(1) Electrical energy can be stored in two plates separated by an insulating medium. Such a device is called a capacitor, or a condenser, and its ability to store electrical energy is termed capacitance. It is measured in farads.

电能能被储存在由一绝缘介质隔开的两块金属板内。这样的装置被称为电容器，其储存电能的能力被称为电容，电容的测量单位是法拉。

(2) We can store electrical energy in two metal plates separated by an insulating medium. We call such a device a capacitor, or a condenser, and its ability to store electrical energy capacitance. It is measured in farads.

我们能把电能储存在由一绝缘介质隔开的两块金属板内。我们把这样装置称为电容器，把其储存电能的能力称为电容，我们以法拉为电容的测量单位。

The subjects of each sentence in (1) are:

Electrical energy

Such a device

Its ability to store electrical energy

It (Capacitance)

They all include more information. Being placed at the beginning of the sentence, they are very striking. Four subjects are different totally, which prevents from dull repeating. The front and back are consistent, smooth naturally. It shows that passive structure can accept the succinct and objective result.

The Use of Post Modifiers

It is one of the characteristics of scientific literature to use a lot of post modifiers, too. The common modifiers are prepositional phrase, adjective or adjective phrase, adverbial word, individual participle and attributive clause, etc.

Prepositional Phrase

The forces due to friction are called frictional forces.

由于摩擦而产生的力称为摩擦力。

Adjective or Adjective Phrase

Antarctica is a continent very difficult to reach.

南极洲是一块很难到达的大陆。

Since 1930 dozens of objects smaller than atoms have been discovered.

1930 年以来，发现了几十种比原子还小的物质。

Semiconductors have properties different from those of conductors and insulators.

半导体具有不同于导体和绝缘体的特性。

Adverbial Word

The air outside pressed the side in.

外面的空气将桶壁压得凹进去了。

The force upward equals the force downward so that the balloon stays at the level.

向上的力与向下的力相等，所以气球就保持在这一高度。

Individual Participle

A material subjected to a tensile or compressive load of sufficient magnitude will deform at first elastically and then plastically.

受一定大小的拉伸载荷或压缩载荷的材料，首先发生弹性变形，然后发生塑性变形。

Attributive Clause

Anyone who thinks that rational knowledge need not be derived from perceptual knowledge is an idealist.

如果认为理性知识不必由感性知识得来，那他就是一个唯心主义者。

It was not until sixty years ago that a method of extracting aluminum ore was found which could lead to a cheap large-scale process.

直到 60 年前人们才找到开采铝矿的方法，从而使低成本、大规模冶炼金属铝成为可能。

Liquids, which contain no free electrons, are poor conductors of heat.

各种液体，由于不含有自由电子，是热的不良导体。

Translation Techniques of Scientific Literature

Translation is an extremely complicated social psychological phenomenon in the information interchange course. The knowledge of the language is an interpreter's foundation. In addition, it is a complicated psychological cognition process involving reasoning, judging, analyzing and synthesizing etc. The translation methods and skills are the universal laws that interpreter summarized according to the characteristics of two languages in the long-term practice. They can guide us on translating practice and help us deal with various kinds of language phenomena in translating course more conscientiously and agilely.

Subordinate Clause and Diction

In scientific literature, three kinds of subordinate clauses are used frequently. They are nominal, attributive and adverbial subordinate clause. The adverbial subordinate clause can be sub-divided to an adverbial of time, place, cause, comparison, proportion, condition, concession, purpose, result etc. Before the translation, it is important and necessary to distinguish which is the main clause and which is the subordinate clause.

For the predicative, objective subordinate clauses or nominal subordinate clauses guiding by such pronouns as what, who, whether, that, where, why, how, when. They can be translated in the original order.

(1) That like charges repel but opposite charges attract is one of the fundamental laws of electricity.

同性电荷相斥、异性电荷相吸是电学的基本定律之一。

(2) The great advantage of electronic computers is that they do many operations in an incredibly short time.

电子计算机最大的优点是，它们能在短到令人难以置信的时间内进行大量运算。

(3) Wilmut maintains that cloning animals has tremendous potential for helping people.

威尔姆坚持认为克隆动物在为人类造福方面有着巨大的潜力。

In the structure of forms subject it + predicate + subordinate clauses guiding by that (whether), it should be translated in it original order.

(1) It isn't surprising that much of today's medical software is designed to help practitioners handle the information over load.

毫不奇怪，今天设计的许多医疗软件都用来辅助医生应付信息量超负荷的问题。

(2) It is estimated that the new synergy between computers and Net technology will have significant influence on the industry of the future.

据预测，计算机和网络技术之间的新的结合将会对未来工业产生巨大的影响。

Amplification and Omission

Because of the difference between English and expression way of Chinese, English original texts and Chinese translations are not one-one relations definitely. Some words are indispensable in English, but it is surplus to translate out word by word. On the other hand, when translating English sentences into Chinese, it is necessary to add some words which are absent in English original text. These kinds of amplification and omission translation methods are indispensable ways to express the original text accurately and fluently.

It should be noted, amplification method doesn't mean to create something out of nothing again, omission method doesn't mean to delete sth. willfully. Whether it is appropriate of using amplification

and omission translation methods, it depends on the skilled degree of the translator for the original language and translation language. That is to say, it depends on the depth in understanding to English original text and accuracy in using Chinese expression.

(1) You have seen how water expands when it is heated and contracts when it is cooled.

你已经看到水受热时怎样膨胀，冷却时又怎样收缩。

(2) Now human being have not yet progressed as to be able to make an element by combining protons, neutrons and electrons.

目前，人类的技术还没有进展到能把质子、中子、电子三者化合成一个元素的地步。

(3) A square has four equal sides.

正方形四边相等。

(4) Technology is the application of scientific method and knowledge to industry to satisfy our material needs and wants.

技术就是在工业上应用科学方法和科学知识以满足我们物质上的需求。

Negation Methods

Both English and Chinese can express the same concept from front or the reverse side. In practical English-Chinese translation, in order to deal with the expression difficulty and to make the translation more clearly and coherently, sometimes we can translate the original text from the opposite expressed side.

Negation method is an important translation method. It is used widely in practical translation.

(1) One body never exerts a force upon another without the second reacting against the first.

一个物体对另一个物体施加作用力必然会受到另外一个物体的反作用力。

(2) Metal do not melt until heated to a definite temperature.

金属加热到一定温度才会熔化。

(3) Before too many years have passed, corner gasoline stations may be replaced by ammonia or methanol stations or battery-recharging terminals.

不用多少年，街头的加油站就可能被氨或甲醇供应站或者是蓄电池充电站所代替。

(4) The average speed of all molecules remains the same so long as the temperature is constant.

只要温度不变，全部分子的平均速度也就不变。

Division Methods for Long Sentences

It is very common to express complicated concepts with long sentence in English. The appearance of long sentences in scientific literature is extremely frequent. Generally speaking, there are three respects to lead to the fact of long sentences. ① There are too many modifiers. ② There are too many side by side structures. ③ There are many structure levels of language. Chinese is different, which often uses several short sentences and does the well arranged narration. So, while translating from English to Chinese, special attention should be paid to the difference between English and Chinese, resolve the long sentence of English, translate into short sentences of Chinese. The following methods are generally used.

1. When the content and narrating level of long English sentence are unanimous basically with Chinese. It can be translated into Chinese according to the order of English original text. For example:

(1) However, fiber systems can carry so many more telephone conversations at the same time

than wire pairs, and can carry them so much farther without amplification or regeneration, that when there are many telephone calls to be carried between points such as switching offices fibers systems are economically attractive.

然而，光导纤维系统在同一时间内传送的电话比普通导线传送的多，它无须放大或者再生就能将之传送到更远的地方，因此，当各地之间（例如中继局之间）要进行大量的通话时，光导纤维系统就会在经济上显得更具吸引力。

(2) In the not-too-distant future, computers will probably give on every aspect of farming—which crops to plant, when and where to plant them, how much and what types of fertilizer to use, and when and where to market the crops.

在不远的将来，计算机也许将为农业提供方方面面的帮助，如种植哪些作物，在什么时候种，在哪些地方种，用哪种肥料，用量多少，以及在什么时候和什么地方销售农产品。

2. The expression order of English long sentences is different from that of Chinese, even totally opposite. So it must be translated from the back of the original text. For example:

(1) Aluminum remained unknown until the nineteenth century, because nowhere in nature is it found free, owing to its always being combined with other elements, most commonly with oxygen, for which it has a strong affinity.

铝总是跟其他元素结合在一起，最普遍的是跟氧结合；因为铝跟氧有很强的亲和力，由于这个原因，在自然界找不到游离状态的铝。所以，铝直到19世纪才被人发现。

(2) This is why the hot water system in a furnace will operate without the use of a water pump, if the pipes are arranged so that the hottest water rises while the coldest water runs down again to the furnace.

如果把管子装成这样，使最热的水上升，而最冷的水流下来后返回锅炉里去，那么锅炉里的热水系统不用水泵就能循环，道理就在于此。

3. Sometimes, if the relation between subject or the main clause and the modifier is very close in English, according to the habit of using more short sentences in Chinese, the subordinate clauses or phrases of sentence can be turned into the sentences while translating. That is to say, to narrate separately in order to make language purpose consistent. Sometimes some words should be added properly.

(1) Lathe sizes range from very little with the length of the bed in several inches to very large ones turning a work many feet in length.

车床尺寸有大有小，小的床身只有几英寸长，大的可以切削几英尺长的工件。

(2) Manufacturing process may be classified as unit production with small quantities and mass production with large numbers of identical parts being produced.

制造过程可以分为单件生产和批量生产。单件生产就是生产少量的零件，批量生产是指生产大量相同的零件。

Inversion

Some inversion structures appear because the grammar needs, some appear because of the need that the structure should be balanced.

(1) Punched on the tapes are numbers of holes which mean binary's 1 for the computer.

带上打着许多孔，这些孔对计算机来说意味着二进位的"1"。

(2) Only after a program is prepared in every detail, can the electronic computer understand the

problem it to solve.

只有在详尽地编织出程序后，电子计算机才能读懂要解的题目。

(3) Never has a machine been so efficient and accurate as the electronic computer.

从来不曾有过像电子计算机那样效率高而又准确的机器。

(4) No sooner has the current started running in one direction than back it comes again.

电流刚开始朝一个方向流动就立即返回。

Notes

1. The architecture of scientific literature advocates preciseness, accurate concepts, strong logic, concise style of writing, outstanding emphases, neat and less changeable sentence type.

句意：科技文献的结构应体现严谨、概念准确、逻辑性强、文风简练、重点突出、句式少有变化等特点。

Exercises

Translate the following sentences into Chinese.

1. One material can be distinguished from another by their physical properties: color, density, specific heat, coefficient of thermal expansion, thermal and electrical conductivity, strength and hardness.

2. Many plastics become soft on heating but harden again when cooled, while others can be heated strongly without apparently changing.

3. It has to be pointed out that one and the same word may have different meanings in different branches of science and technology.

4. But the profligate burning of petroleum products seems a great waste of those useful petrochemicals that could otherwise be turned into lubricant synthetic, and many other useful products.

5. We learn that sodium or any of its compounds produces a spectrum having a bright yellow double line by noticing that there is no such line in the spectrum of light when sodium is not present, but that if the smallest quantity of sodium be thrown into the flame or other sources of light, the bright yellow line instantly appears.

6. Manufacturing process may be classified as unit production with small quantities being made and mass production with large numbers of identical parts being produced.

Chapter 13 Scientific Papers Writing

Writing Methodologies

The course of scientific papers writing, in a broad sense, is the course of scientific research. It also means the whole course from scratching to finishing the papers in a narrow sense. It will be discussed in the narrow sense in this section. The writing course of scientific papers can be divided into the following several steps: choosing a topic, reviewing and collecting information, analyzing the information, summarizing and organizing ideas, working out an outline, writing the first draft, revising the draft and finalizing the paper.

Choosing a Topic

The choosing and confirmation of the topic is the first step in scientific papers writing. The topic should reflect the ideas and viewpoints refined from concrete materials. These ideas and viewpoints are called the theme, which are the soul of the scientific papers. The value of any scientific paper depends upon whether the theme is correct and deep or not. After the final conclusion is worked out from the subject, the theme suitable for scientific papers writing is clearly conspicuous. There are two respects. One is that the conclusion proves the original imagination at first when choosing the topic. Namely the theme has been confirmed at the beginning of choosing the topic. The other is that the conclusion in research denies the initial imagination wholly or partly. Thus a new conclusion is drawn. This is a new theme. In brief, the theme is the embodiment of the writing intention.

There are also some basic principles for choosing a good topic that can present the distillation of the papers.

(1) The topic should be chosen concretely, not trifling and isolated. So the reader can know clearly that what subject is studied.

(2) The difficulty of subject should be moderate. It shouldn't be beyond the writer's ability. The feasibility of the subject should be considered from every side.

(3) The subject should maximize favorable factors and minimize unfavorable ones. Make every effort to fit in with the specialty that the writer studies.

(4) The subject had better not be done by forefathers in this discipline. Or forefathers have done something on it, but still not complete or there is falsehood.

(5) The subject direction is a promising direction.

(6) Based on discipline, those subjects having certain practical value for national economic construction should be chosen emphatically.

Reviewing and Collecting Information

The scientific papers are the articles which state the scientific findings. It is a precondition for scientific papers writing to occupy the materials in details. The essence of scientific research is to find the inherent law of things and announce its true. The laws always exist in a large number of phenomena and the true essence always is contained in the numerous and complicated materials. To occupy the materials exhaustively, it is possible for the researchers to obtain their inherent laws instead of fabricating conclusions in the analysis and research of materials.

From the points of view of writing, the materials are the flesh and blood of the article. They are also the components of the content, and the backing and pillar of the argument of the thesis. If there is only argument with poor materials and insufficient grounds, it can't prove argument clearly, concretely and effectively. The argument will seem unconvincing and unable. The discovery of the problem and the clue of solving the problem always exist in the materials. The more abundant the occupation of the materials is, the clearer the question will be. Only after occupying a great deal of materials, can we make clear what the problems that our forefathers haven't solved or mentioned yet in the range of research topic. If we collect more materials about these questions, we can make clear where the key of the problems is, and then may find the correct answer to the question.

Generally, the collection of materiel must follow the following requirements.

(1) The materials should be absolutely necessary and abundant. Otherwise the theme can't be well displayed because of lacking of essential materials. While writing, the author should catch this kind of materials. Don't adopt the materials having nothing to do with the theme, no matter how difficultly to obtain them. But the materials should be abundant, too. If there is no certain quantity, it is difficult to prove the question clearly, namely so-called "lack of evidence ".

(2) The materials should be true and accurate. The materials should come from the objective reality, namely come from social investigation, production practices and scientific experiments. The scientific papers emphasizes much on scientific. Any untrue or inaccurate materials will make the view lose the credibility and dependability. Thus in order to obtain the true and accurate materials, the research approach, investigation method and experiment scheme should be chose rationally, experiment operation and the collection, dealing with of data should be correct. Try best to use the direct materials and to analyze and check the undirect materials.

(3) The materials should be typical and novel. Namely the materials can reflect the essential characteristics of things. Such materials can make the reason specific and make the description visual. In order to get the typical material, the investigation and research work must be deepened. Otherwise it is difficult to catch the essence of things. Novelty means fresh, not outmoded. To make the material novel, it is the key to do pioneering work to obtain the innovative achievement constantly. Meanwhile, the materials should be selected widely. By analyzing and comparing, choose the new material that can reflect the new development, new achievements, and abandon the outmoded materials.

Materials of scientific and technological thesis, mainly include the scientific and technological monograph, scientific and technological report, scientific and technological digest, patent documentation, technical standard, various kinds of collections of thesis (including periodical thesis, collection of thesis, meeting papers, academic dissertation etc.), research report, laboratory report, experimental technique report, observing notes (includes drawing, photo, recording, video etc.), technical appraisement materials and journal , etc.. There are also some ways to collect above-mentioned relevant materials.

(1) Participate in relevant experiments, test and make investigations to obtain the firsthand materials.

(2) Participate in such relevant meetings as the symposium, technological evaluation meeting, seminar, technological seminar, project discussion, etc. And ask for relevant materials.

(3) Refer to "national new booklist" and go to the bookstore to choose relevant works and periodicals.

(4) Search, read, duplicate and ask for or borrow the useful materials in the library.

(5) Join the information network, obtain relevant materials from the companies concerned or the net explorer in the network, or get relevant scientific and technical literatures from counterpart, friends.

(6) Materials on Internet can act as reference, but should not be regarded as a basis generally.

Analyzing the Information

It is very important to analyze a large number of primitive materials obtained. Clean up the data to form a form. When necessary, draw a figure or take photography in order to keep the true appearance. Reexamine, crosscheck some materials, so as to ensure it is correct.

Centering on the subject and thesis content closely, distinguish and choose the material strictly. See clearly the nature of the material. Distinguish the true from the false, estimate the value of the materials and weigh the function of the materials. Be sure to give up the materials having nothing to do with the thesis, otherwise it will make the content numerous and jumbled. For the materials that are already chosen, they should be pondered deeply and considered repeatedly in order to deepen and improve them in theory.

After distinguishing and choosing, duplicate needed materials by applying the technological means of copying or miniature. Sort out them separately according to nature for use while writing. The arrangement level of the materials should be distinct. It should embody inner link among materials.

Summarizing and Organizing Ideas

As soon as we have decided upon a topic and our purpose in writing about it, we should start to summarize and organize ideas, a step that helps us to think of as many ideas as possible about the topic. The generation of ideas takes different forms. In some cases, if the writer intends to write a fairly short piece on a topic about which she/he has good knowledge, brainstorming can create ideas. In practice, the writer sets aside a period of time to think of ideas for the writing, jotting them down on paper for further consideration. More frequently, however, a certain amount of research work will be necessary to generate ideas, especially when the writer intends to write extensively on an academic topic. Depending on the size of the task, this step may take anywhere from a few of minutes to several months or even longer.

The process of research achieves three purposes. One is to help you determine what others have already done in the area. The second is to acquire the basic methods of research. The third is to generate ideas. By the end of the research stage, you will have clearer ideas as well as a rough plan for your paper.

Working Out an Outline

A clear outline can help us to organize the materials logically. It should be a framework that lists the ideas in the order in which you will discuss them. Each part should be headed by a short chapter title, which is followed by a topic sentence. This framework will serve as the roadmap for the next stage.

There are two types of outlines: sentence outlines and topic outlines. In a sentence line each entry is a complete sentence. In a topic outline each entry is a phrase or a single word. When making a sentence outline, the writer has to think out each entry to a greater degree than for the topical form. As a result, the possibility of ambiguity and vagueness is lessened. As the sentence outline takes a longer time, you may write instead a topic outline, which more practical as a guide for writing. The correct outline form follows 5 specific rules:

(1) Place a title at the very beginning.

(2) Write out the central idea for the whole outline.

(3) Use Roman numerals Ⅰ, Ⅱ, Ⅲ, etc. to designate the main ideas in the paragraphs which are used to prove the central idea of the whole passage.

(4) Designate supporting ideas and expanding ideas at different levels in descending order, first, by capital letters (A, B, C, etc.), next, by Arabic numerals (1, 2, 3, etc.), then small letters (a, b, c, etc.), then Arabic numerals in parentheses ((1), (2), (3), etc.), followed by small letters in parentheses ((a), (b), (c), etc.).

(5) Be sure that there are two or more main ideas, two or more supporting ideas and so on. Because whenever you divide anything up, the minimum number of parts is two.

Writing the First Draft

Now comes the time to write a first draft. Writers generally try to follow the outline closely while developing each idea fully. If new ideas appear at this stage, they are jotted down for later incorporation and development, if there are too many new ideas for the framework to contain, one needs to regress to restart with a fresh outline. This is the stage where a change of mind most often occurs, and therefore an open mind for new ideas is important.

The writer would also concentrate more on the development of ideas at this stage than the correctness of grammar, spelling, and other surface features, a task for the editing and proofreading stage. After finishing the first draft, the writer usually sets it aside for a time, a number of hours or days according to time limits, this creates a distance between the writer and the writing so that when the time comes to read the draft, it is done more objectively and cool-headedly.

Revising the Draft and Finalizing the Paper

Revision is an extremely important step in the process of writing. No writer expects to write clearly and satisfactorily the first time round. Good writing is the result of repeated revision. Indeed, it is usually the good writer who finds that many parts of the first draft are unsatisfying.

Some of the ideas may not have been adequately developed, while other ideas may have failed to support the thesis. A number of new ideas illustrating the arguments even better may have appeared. In the process of revision, the writer may also return to do more research, thinking, and organizing. As a result, by the time the second draft is finished, it is usually quite different in terms of organization and presentation of ideas. Even the central idea may have been adjusted because of new materials and facts gathered from the research. The time must come when you consider the work to be as good as you can make it.

Writing, as we see it, is a complex and recursive process. The different stages of research, planning, drafting, revising, and editing are all important towards the target of a good final product.

General Structures and Formats

The general structure of a thesis is a unified whole. From the beginning, the middle to the ending, it should be consistent, well arranged, and logical. There is an intact composition. The basic pattern of the general structure of scientific thesis consists of three main parts: The preface, body and conclusion.

(1) The preface proposes the cause and importance of research subject. It also introduces the basic situation in this field.

(2) The body should regard proposition as the key link. The argument should be clear and conclusive. The scientific paper is refined more concisely in order to express the main, most excellent and creative content in research work. The view must be distinct and proved by the facts. But it is unsuitable to enumerate too many facts. If the documents materials or others' are quoted, they should be marked in the footnote or the list of references to show the sources.

(3) The conclusion puts forward the result briefly, which problems are solved and which questions remain unsolved for further discuss later.

A summary gives the main ideas, major subtopics and important details of a body of materials. Here are the steps you should follow to make a good summary of an article.

1) Preview the article by taking a quick look at the following: title, subtitle, the first paragraph and a few paragraphs at the end, and other items such as heads or subheads, pictures, charts or diagrams, words or phrases set off in italic type or boldface print, etc.

2) Read through the article for general statements and for details or examples that support the statements and mark them.

3) Reread more carefully the areas you have identified as the most important. Also, focus on other key points you may have missed in your first reading.

4) Take notes on the materials.

Analyses of Typical Issues

After the scientific and technological thesis is issued out, its purpose lies in exchanging mainly. So it should be easy to understand. Generally, there are three main issues in scientific papers.

1. Logic issues

The so-called logic, that is to say, the reason must be explained openly and it should not be upside down and self-contradictorily. Without logical structure, the readers cannot understand the thesis well. So the article will be lack of persuasion.

2. Normal form issues

3. Grammar issues

Grammar issues are the most common issues in scientific papers. The following are some common grammar issues.

(1) There is some improper matching. It is the common trouble in the language to collocate improperly.

(2) The composition is incomplete. The sentence lacks of some essential composition. The most common one is lacking the subject or predicate.

(3) The composition position is upside-down. Various kinds of additional compositions of the sentence must be placed on the suitable positions according to the nationality customs and grammar requirements. Otherwise it will be ambiguous.

(4) Several words or phrases with different kinds of nature stand side by side acting as the same composition of the sentence. It will often make the whole meaning ambiguous.

(5) The subject is substituted for another surreptitiously. A same thing is explained with several sentences, but there is no consistent subject.

(6) The morphological feature is misapplied. There are different usages of different kind of words in the sentence. If misapplying the class-A word as the class-B word, that is the morphological feature is misapplied.

(7) What a pronoun refers to must be clear. Otherwise it is very easy to cause misunderstanding.

(8) The languages of various nationalities are different and the word orders are different, too.

Exercises

Correct the mistakes in the following sentences.

1. Make our cities greener is important. Plant trees and flowers is the best measure to obtain the goal.

2. My summer's work proved not only interesting but I also learned much from it.

3. She has fallen in love with him not because he is handsome but that he is diligent.

4. The increasing use of chemical obstacles in agriculture also makes pollution.

5. The old man returning home after eight years' absence to find that all the neighbors he had known were no longer there.

Writing task

6. Choose a topic that you are interested in, and go to the library in your university to collect relevant information, and then write an outline of the scientific paper about the topic.

Appendix

Appendix A Websites

1 http: // www. iie. org/
2 http: // www. iienet. org
3 http: //www. asme. org/
4 http: //www. sme. org/
5 http: //www. apics. org/
6 http: //www. asq. org/
7 http: //www. cmes. org/
8 http: //www. myie. org/
9 http: //www. ie. tsinghua. edu. cn/
10 http: //www. chinaie. info/
11 http: //www. ie56. com/
12 http: //www. iecn. org/
13 http: //www. manufacturing. net
14 http: // www. advancedmanufacturing. com
15 http: //www. manufacturingnews. com
16 http: // www. manufacturingweek. com
17 http: // www. manufacturingtalk. com
18 http: // www. modine. com
19 http: // www. manufacturingcenter. com
20 http: // www. china-machine. com. cn
21 http: // www. cims. tsinghua. edu. cn
22 http: //www. e-works. net. cn
23 http: //www. unep. org
24 http: //rpm. xjtu. edu. cn
25 http: //www. ergonomics. org. uk/
26 http: //www. ergonomics. org/
27 http: //www. ergonomics. ucla. edu/
28 http: //www. hfes. org/web/Default. aspx
29 http: //www. ergonomics. com. au/
30 http: //www. apa. org
31 http: //psych. wlu. edu

32 http: //www. cqm. org/

33 http: //www. americanquality. com/

34 http: //www. qmi. com/

35 http: //www. isixsigma. com/

36 http: //www. quality. co. uk/

37 http: //www. shrm. org/

38 http: //www. dhrm. state. va. us

39 http: //www. nhrma. org/

40 http: //www. hrma. org/

41 http: //www. hr-guide. com/

42 http: //www. conferenceboard. ca/humanresource/default. htm

43 http: //www. nlc. gov. cn

44 http: //www. cnki. net

45 http: //www. cqvip. com

46 http: //www. wanfangdata. com. cn

47 http: //www. baidu. com

48 http: //www. google. com

49 http: //lib. nju. edu. cn/nju_ resource. htm

50 http: //www. eas. pdx. edu/systems/lgewelcome/module1. htm

51 http: //research. hq. nasa. gov/code_ y/nra/current/AO-99-OES-01/ preproposal /intro/ index. htm

52 http: //www. manufacturing. net/scm/

53 http: //www. cio. com/research/scm/edit/012202_ scm. html

54 http: //logistics. about. com/

55 http: //www. mhhe. com/business/mis

56 http: //jmis. bentley. edu/

57 http: //www. isworld. org/

58 http: //www. tongji. edu. cn/ ~yangdy/technology/APTS/CHAP2. HTM

59 http: //isg. urv. es/publicity/masters/sample/techniques. html

60 http: //www. fi. muni. cz/usr/wong/teaching/mt/notes/mt. html. iso-8859-1

61 http: //www. doc. ic. ac. uk/ ~nd/surprise _97/journal/vol2/hks/lan_ trans. html

62 http: //mason. gmu. edu/ ~arichar6/logic. htm

63 http: //www. columbia. edu/cu/biology/ug/research/paper. html

64 http: //sportsci. org/jour/9901/wghstyle. html

65 http: //www. ag. iastate. edu/aginfo/checklist. html

66 http: //www. techdirection. com

67 http: //www. zjie. com/

68 http: //www. sou23. com/info/gongyegongcheng. htm

69 http: //www. gdpx. com. cn/web/bbs/

70 http: //netec. wustl. edu/WoPEc/data/Papers/dgrkubcen199870. html)

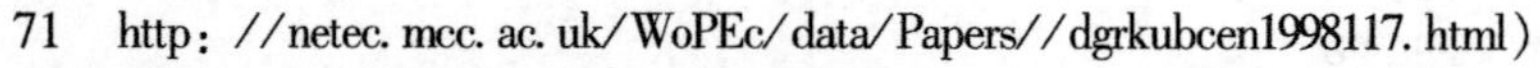
71 http://netec.mcc.ac.uk/WoPEc/data/Papers//dgrkubcen1998117.html)

72 http://boss.3726.cn/subject/gygc/

73 http://www.techdirection.com

Appendix B Professional Words and Expressions

1. a code of ethics 道德规范
2. absolute accuracy 绝对精度
3. abstract *n.* 摘要
4. accountant *n.* 会计（员），会计师
5. action learning 行动学习
6. adherence *n.* 忠诚
7. afterwards *adv.* 之后，以后，后来
8. AGV（Automated Guided Vehicles）自动导航小车
9. aligning results 校正结果
10. allocation *n.* 配合，分配；分配额
11. alternation ranking method 交替排序法
12. AM（Agile Manufacturing）敏捷制造
13. analyst *n.* 分析者；善于分析者；分解者
14. anatomical *adj.* 解剖的，解剖学的
15. annual bonus 年终分红
16. anthropometric *adj.* 人体测量的
17. anthropometry *n.* 人体测量学
18. appeal *n.* 申诉，请求，呼吁，上诉，要求
19. application form 工作申请表
20. appraisal interview 评价面试
21. appreciation *n.* 正确评价
22. aptitude *n.* 资质
23. arbitration *n.* 仲裁
24. architecture *n.* 结构，构造
25. arena *n.* 舞台，竞技场
26. assembly line 装配线
27. assurance 确信，断言，保证
28. attack *vt.* 动手处理（某事）；攻击，抨击
29. attainment *n.* 到达
30. attendance incentive plan 参与式激励计划
31. audit *n.* 审计，稽核，查账
32. auditor *n.* 审计员，核数师
33. authority *n.* 职权
34. awkward *adj.* 难使用的，笨拙的
35. backdate *vt.* 回溯
36. batch production 批量生产
37. be closely intertwined 紧密融合在一起
38. be prone to 倾向于……
39. behavior modeling 行为模拟
40. behaviorally anchored rating scale（bars）行为锚定等级评价法
41. below contributor 贡献较小的人员
42. benchmark job 基准职位
43. benefit *n.* 福利
44. bias *n.* 个人偏见
45. biomechanical *n.* [生] 生物力学
46. blanking *n.* 消除，切断，空白
47. Boolean logic 布尔逻辑
48. boycott 联合抵制
49. broach *vt.* 拉削
50. budget analyst 预算分析师
51. bumping/layoff procedures 工作替换/临时解雇程序
52. bdurnout *n.* 耗竭
53. bushing *n.* 衬套
54. business management 企业管理
55. CAE（Computer-Aided Engineering）计算机辅助工程
56. caliper *n.* *vt.* 卡尺，卡钳；用卡钳测量
57. candidate-order error 候选人次序错误
58. canons of ethics 道德标准
59. capital accumulation program 资本积累方案
60. CAPP（Computer-Aided Process Planning）计算机辅助工艺规划
61. career anchors 职业锚

62. career cycle 职业周期
63. career group 事业组
64. career group description 事业组描述
65. career planning and development 职业规划与职业发展
66. case study method 案例研究方法
67. cell production 单元生产
68. cellular layout 单元式布局
69. central tendency 居中趋势
70. checksheet *n.* 检测工作表，记录表，调查表
71. chuck *n.* 卡盘
72. CIM（Computer Integrated Manufacturing）计算机集成制造
73. citation *n.* 传讯
74. Civil Rights Act 民权法
75. classification（or grading）method 归类（或分级）法
76. clause *n.* 从句
77. CNC（Computerized Numerical Control）计算机数字控制
78. code design 代码设计
79. cognitive *adj.* 认知的，认识的，有感知的
80. coil feeder 卷料进料装置
81. collective bargaining 集体谈判
82. colloquia *n.* 座谈会（colloquium 的复数）
83. commonality *n.* 共性
84. comparable worth 可比价值
85. compensable factor 报酬因素
86. compromise *n.* 妥协，折衷
    v. 妥协，折衷，危及……的安全
87. compulsory *adj.* 必修的，强制的
88. computerized forecast 计算机化预测
89. confidence *n.* 置信度
90. confidentiality *n.* 机密性
91. connotation *n.* 内涵
92. content validity 内容效度
93. continuous timing method 连续记时法
94. contributor 贡献者，捐助者，投稿者
95. controversy *n.* 论战
96. convoluted *adj.* 复杂的，旋绕的
97. CONWIP control 定量在制品控制
98. core responsibility 核心职责，核心责任
99. corkscrew *n.* 螺丝锥
100. criterion *n.*（判断的）标准，准据，规范
101. criterion validity 效标效度
102. critical incident method 关键事件法
103. current flow 电流
104. curriculum *n.* 课程（pl. curricula）
105. C-VARWIP control 循环变量在制品控制
106. cybernetics *n.* 控制论
107. dashboard *n.* 汽车等的仪表板
108. Data Structure Diagram（DSD）数据结构图
109. Davis-Bacon Act（DBA）戴维斯-佩根法案
110. day-to-day-collective bargaining 日常集体谈判
111. deadlock *n.* 停顿，思考
112. decline stage 下降阶段
113. deferred profit-sharing plan 延期利润分享计划
114. defined benefit 固定福利
115. defined contribution 固定缴款
116. degradation *n.* 退化
117. demographics *n.* 人口统计学
118. department of labor job analysis 劳工部工作分析法
119. diligently and persistently 坚持不懈地
120. discipline *n.* 纪律；学科
121. disengage *v.* 脱离，释放
122. dismissal *n.* 解雇；开除
123. dispersion *n.* [数] 离差，差量，平均值，离中趋势
124. dissipate *v.* 驱散
125. distract *v.* 转移
126. distraction *n.* 娱乐，分心，分心的事物
127. doctoral dissertation 博士论文
128. domain *n.* 领域，范围
129. downsizing *n.* 精简
130. drill press 钻床

131. dumping *n.* 倾销
132. early retirement window 提前退休窗口
133. economic batch quantity 经济批量
134. economic strike 经济罢工
135. Edgar Schein 艾德加·施恩
136. e-manufacturing 网络化制造
137. empathy *n.* 移情作用
138. employee compensation 职员报酬
139. employee orientation 雇员上岗引导
140. Employee Retirement Income Security Act (ERISA) 雇员退休收入保障法案
141. employee services benefits 雇员服务福利
142. Employee Stock Ownership Plan (ESOP) 雇员持股计划
143. encyclopedia *n.* 百科全书
144. Enterprise Resource Planning (ERP) 企业资源计划
145. Entity Relationship Model (E-R Model) 实体关系模型
146. Equal Pay Act 公平工资法
147. equitable *adj.* 公平的；公正的
148. ergonomics *n.* 人类工程学，生物工程学，工效学
149. ergonomist *n.* 工作学者，生物工程学者
150. essence *n.* 本质，实质
151. establishment stage 确立阶段
152. exemplary *adj.* 可仿效的，可做模范的
153. exit interviews 离职面谈
154. expectancy chart 期望图表
155. experience economy 体验经济
156. experimentation *n.* 实验
157. exploration stage 探索阶段
158. extraordinary contributor 特别贡献者，杰出贡献者
159. extrapolate *v.* 推断
160. facing tool 端面车刀
161. fact-finder 调查
162. fair days work 公平日工作
163. fair labor standards act 公平劳动标准法案
164. FEA (Finite Element Analysis) 有限元分析
165. feedback *n.* 反馈
166. fin *n.* 鳍，鱼翅；鳍状物
167. finished goods 成品
168. flex place 弹性工作地点
169. flexible benefits programs 弹性福利计划
170. flextime 弹性工作时间
171. flow diagram 线路图
172. flow process chart 流程程序图
173. FMS (Flexible Manufacturing System) 柔性制造系统
174. forced distribution method 强制分布法
175. foreign element 外来单元
176. forklif *n.* *vt.* 铲车，堆高机；用铲车搬运
177. fortification *n.* 防御工事，要塞，筑城术
178. foundry *n.* 铸造，翻砂，铸工厂，玻璃厂，铸造厂
179. four-day workweek 每周4天工作制
180. frustration *n.* 挫败，挫折，受挫
181. functional control 职能控制
182. functional job analysis 功能性工作分析法
183. gain sharing 收益分享
184. gang process analysis 联合操作分析
185. gang process chart 联合程序图
186. general economic conditions 一般经济状况
187. GNP (Gross National Product) 国民生产总值
188. golden offerings 高龄给付
189. good faith bargaining 真诚的谈判
190. grade *n.* 等级
191. grade description 等级说明书
192. graphic rating scale 图尺度评价法
193. grapple with 设法解决
194. gravitation *n.* 万有引力
195. grid training 方格训练
196. grievance *n.* 抱怨
197. grievance *n.* 不满；不平；冤情；抱怨；牢骚
198. grievance procedure 抱怨程序

199. grip *vt*. 紧握，紧夹 *vi*. 抓住 *n*. 掌握，控制，把手
200. group life insurance 团体人寿保险
201. group pension plan 团体退休金计划
202. growth stage 成长阶段
203. GT（Group Technology）成组技术
204. guarantee corporation 担保公司
205. guaranteed fair treatment 有保证的公平对待
206. guaranteed piecework plan 有保障的计件工资制
207. halo effect 晕轮效应
208. hand in hand 合作
209. hand over 交出，移交
210. handbook *n*. 手册
211. health maintenance organization(HMO) 健康维持组织
212. heterogeneity *n*. 异种，异质，不同成分，多种多样
213. Hierarchy Plus Input，Process，and Output（HPIPO）层次的输入，处理，输出图
214. histogram *n*. 柱状图，直方图
215. homeostasis *n*. 动态平衡
216. households *n*. 家庭，一家人
217. hybrid layout 混合式布局
218. hypothenuse *n*. 斜边
219. idlesse *n*. 空闲
220. illegal bargaining 非法谈判项目
221. impasse *n*. 僵持
222. implied authority 隐含职权
223. in terms of 就……而论；在……方面
224. incentive plan 激励计划
225. index *n*. 索引
226. Individual Retirement Account（IRA）个人退休账户
227. industrial economy 工业经济
228. industrial engineering 工业工程
229. industrial robot 工业机器人
230. industrialization *n*. 工业化，产业化
231. infrastructure *n*. 基本设施，基础设施
232. in-house development center 企业内部开发中心
233. insubordination *n*. 不服从
234. insurance benefits 保险福利
235. intangibility *n*. 无形，不能把握，不可解
236. integrated equipment 集成设备，综合设备
237. intensity *n*. 强烈，剧烈，强度，亮度
238. internship *n*. 实习医师
239. intervention *n*. 干涉
240. interview *n*. 谈话；面谈
241. intuitive *adj*. 直觉的
242. inventory 详细目录，存货，财产清册，总量
243. iterative *adj*. 交互的
244. Jackson Design Methodology（JDM）杰克森设计方法
245. jamming 干扰，人为干扰
246. JIT（Just-In-Time）准时工作制
247. job analysis 工作分析
248. job description 工作描述
249. job evaluation 职位评价
250. Job Instruction Training（JIT）工作指导培训
251. job posting 工作公告
252. job rotation 工作轮换
253. job sharing 工作分组
254. Job specifications 工作说明书
255. John Holland 约翰·霍兰德
256. Joint Application Development（JAD）联合应用开发
257. junior board 初级董事会
258. kanban 看板
259. keyway *n*. 键槽
260. keyword *n*. 关键词
261. kit *n*. 成套工具（或物件等）；工具箱
262. labor dissension 劳动纠纷
263. lathe *n*. 车床
264. lawnmower *n*. 剪草机
265. layoff *n*. 临时解雇，下岗
266. lead time 提前期，生产周期

267. leader attach training 领导者匹配训练
268. Lean Manufacturing 精益生产
269. level indicator 级别指示
270. lever *n.* 控制杆
271. life cycle 寿命周期
272. lifetime employment without guarantees 无担保终身雇用
273. line manager 一线管理者
274. literature search/retrieval 文献检索
275. local market conditions 地方劳动力市场
276. lockout *n.* 停工
277. long hand 长针
278. LP (Lean Production) 精益生产
279. machine cell 机器单元
280. maintenance stage 维持阶段
281. management assessment center 管理评价中心
282. Management By Objectives (MBO) 目标管理法
283. management game 管理竞赛
284. management grid 管理方格训练
285. Management Information Systems(MIS) 管理信息系统
286. management process 管理过程
287. mandatory bargaining 强制谈判项目
288. manhole *n.*（锅炉，下水道供人出入检修用的）人孔，检修孔
289. manipulation *n.* 处理，操作，操纵
290. man-machine chart 人机程序图
291. Manufacturing Resource Planning(MRP Ⅱ) 制造资料计划
292. manufacturing system 制造系统
293. mass service 批量服务
294. Material Requirements Planning(MRP) 物料需求计划
295. matrix *n.* 矩阵
296. mediation *n.* 调解
297. merit pay 绩效工资
298. merit raise 绩效加薪
299. mess up 搞乱
300. method study 方法研究
301. Methods Time Measurement 时间测定方法
302. micrometer *n.* 千分尺
303. mid career crisis sub-stage 中期职业危机阶段
304. millennium *n.* 太平盛世，一千年
305. milling machine 铣床
306. modular arrangement of predetermined motion time standard 预定动作时间标准模值排序法
307. module *n.* 模块，组件
308. more than over 更普遍地
309. motion analysis 动作分析
310. musket *n.* 步枪
311. need improvement 需求改进
312. nondirective interview 非定向面试
313. nonmanufacturing 非制造类（企业）
314. notation *n.* 符号
315. notch *n.* 刻痕，凹口 *vt.* 在……上刻痕
316. nut *n.* 螺母
317. Object-Oriented Programming (OOP) 面向对象编程
318. occupational family 职业族
319. occupational market conditions 职业市场状况
320. occupational orientation 职业倾向
321. Occupational Safety and Health Act 职业安全与健康法案
322. Occupational Safety and Health Administration (OSHA) 职业安全与健康管理局
323. occupational skills 职业技能
324. OMT (Object Modeling Technique) 对象造型技术
325. one after the other 一个接一个
326. On-the-Job Training (OJT) 在职培训
327. open-door 敞开门户
328. operation analysis 操作分析
329. operation planning 作业计划
330. operation process chart 工艺程序图
331. operations research 运筹学

332. operations scheduling 作业调度
333. opinion survey 意见调查
334. optimization *n.* 最佳化，优化
335. Organization Development（OD）组织发展
336. organizational objective 组织目标
337. organization's preferred goals 组织的优先目标
338. orientation *n.* 方向，方位，定位
339. out of line 不一致，不协调
340. outpace *v.* 超过
341. outplacement counseling 向外安置顾问
342. paired comparison method 配对比较法
343. panel interview 小组面试
344. part family 产品族
345. participant diary/logs 现场工人日记/日志
346. participation 分享，参与
347. parts feeder 送料器，定向料斗
348. patent *n.* 专利
349. pay band 工资段
350. pay grade 工资等级
351. pension benefit 退休金福利
352. pension plan 退休金计划
353. people-first values 以人为本的价值观
354. percentile *n.* 百分点，百分位
355. performance analysis 工作绩效分析
356. performance appraisal 绩效评价，绩效评估
357. performance appraisal interview 工作绩效评价面谈
358. performance development plan 绩效发展计划
359. performance evaluation 绩效评价，绩效评估
360. performance gap 绩效差距
361. performance management 绩效管理
362. perishability *n.* 易腐烂性，易朽性
363. personalization *n.* 个性化
364. personnel（or human resource）management 人事（或人力资源）管理
365. personnel *n.* 人员，职员
366. personnel replacement charts 人事调配图
367. pertinent *adj.* 恰当的，贴切的；中肯的有关的，相关的
368. phase *n.* 阶段，时期［（+in/of）］
369. physiological *adj.* 生理学的
370. physiology *n.* 生理学
371. piecework *n.* 计件
372. Plant Closing Law 工厂关闭法
373. plating *n.* 电镀
374. plug gauge 塞规
375. plywood *n.* 夹板
376. policy *n.* 政策
377. Position Analysis Questionnaire(PAQ) 职位分析问卷
378. position identification information 职位识别信息
379. position replacement card 职位调配卡
380. post-it 便利贴
381. predetermined time system 预定时间系统（标准）
382. predicative *adj.* 谓语的；表语的
383. preferred *adj.* 首选的
384. prefix *n.* 字首，前缀
385. Pregnancy Discrimination Act 怀孕歧视法案
386. prerequisite *adj.* 首先具备的
387. prevailing *adj.* 主要的
388. preventive maintenance 定期检修
389. probationary period 试用期
390. proceeding *n.* 会议录
391. process analysis 程序分析
392. process layout 工艺式布局
393. procurement *n.* 采购，获得
394. product layout 品式布局
395. production line 生产线，流水线
396. production planning 生产计划
397. production process 生产流程
398. production scheduling 生产调度
399. production sequence 生产顺序
400. professionalism *n.* 专业化

401. profit-sharing plan 利润分享计划
402. programmed learning 程序化学习
403. proponent *n*. 建议者，支持者
404. prototype *n*. 原型
405. prudence *n*. 思考
406. PSL（Problem Statement Language）问题描述语言
407. PSRO（Professional Standards Review Organization）（美国）职业标准审查组织
408. psychology *n*. 心理学，心理状态
409. pushchair *n*. 折叠式婴儿车
410. Pythagorean theorem 勾股定理
411. qualification inventory 资格数据库
412. quality circle 质量圈
413. quest for 追求，探索
414. ramp *n*. 斜坡，坡道
415. ranking method 排序法
416. Rapid Application Development（RAD）快速应用开发
417. rate ranges 工资率系列
418. rate *vt*. 评比，评估
419. rating earned 获得的等级
420. ratio analysis 比率分析
421. rationalizing *n*. 合理化
422. raw *adj*. 生的，未加工的 *n*. 擦伤外 *vt*. 擦伤
423. readout *n*. 读数
424. reality shock 现实冲击
425. recall *n*.（计算机）信息检索（能力）
426. rectangle *n*. 矩形，长方形
427. recur *v*. 再发生，复发
428. regression *n*. 衰退
429. relative accuracy 相对精度
430. reliability *n*. 可靠性，信度
431. render *v*. 给予
432. replenishment 补充，补给
433. requirements forecasting 需求预测
434. reserve *n*. 储藏量
435. resolution *n*. 分辨率
436. responsiveness *n*. 响应性，易起反应
437. retirement benefits 退休福利
438. retirement counseling 退休前咨询
439. retirement *n*. 退休
440. revitalization *n*. 复兴，振兴
441. rigor *n*. 严密，精确
442. rings of defense 保护圈
443. role *n*. 角色，任务
444. role playing 角色扮演
445. Rough Cut Capacity Planning（RCCP）粗能力计划
446. routing *n*. 工艺路线
447. salary reference title 工资证明书名称
448. salary survey 薪资调查
449. sales order processing 销售订单处理
450. savings plan 储蓄计划
451. Scallion plan 斯坎伦计划
452. scatter plot 散点分析
453. Science Journal 科学杂志
454. science-technology literature search 科技文献检索
455. scientific management 科学管理
456. scientific papers writing 科技论文写作
457. self directed team 自我指导工作小组
458. self-actualization 自我实现
459. Semantic Data Model（SDM）语义数据模型
460. semi *n*. 半挂车
461. sense of worth 价值观
462. sensitivity training 敏感性训练
463. serialized interview 系列化面试
464. service customization 定制服务
465. service industry 服务行业
466. service management 服务管理
467. service quality 服务质量
468. service systems 服务系统
469. severance pay 离职金
470. sick leave 病假
471. sift through 筛选
472. simultaneity 同时发生，同时
473. situational interview 情境面试

474. skip-level interview 越级谈话
475. small hand 短针
476. snapback method 归零法
477. snapback *n.* 归零
478. social security 社会保障
479. Speak up! 讲出来!
480. special assignment 特殊任务
481. special award 特殊奖励
482. special management development techniques 特殊的管理开发技术
483. spiral *n.* 螺旋（形）；螺线；蜷线
484. stabilization sub-stage 稳定阶段
485. staff（service）function 职员（服务）功能
486. standard hour plan 标准工时工资
487. stature *n.* 身高，身材；（精神、道德等的）高度
488. stock option 股票期权
489. stopwatch *n.* 计秒表，秒表
490. straight piecework 直接计件制
491. strategic plan 战略规划
492. stress interview 压力面试
493. strictness/leniency 偏紧/偏松
494. strikes *n.* 罢工
495. structured interview 结构化面试
496. subject *n.* 主题
497. substandard performance 不达标绩效，标准以下的绩效
498. succession planning 接班计划
499. supplement pay benefits 补充报酬福利
500. supplemental unemployment benefits 补充失业福利
501. survey feedback 调查反馈
502. survey item 调查项目
503. sympathy strike 同情罢工
504. synchrony *n.* 同步性
505. system I 组织体系 I
506. tache *n.* 环节
507. tangible *adj.* 明确的，可见的，切实的
508. tariffs and quotas 关税和配额
509. task analysis 任务分析
510. team building 团队建设
511. team or group 班组
512. termination at will 随意终止
513. termination *n.* 解雇，终止
514. the instant observation method 瞬间观察法
515. theory X　X 理论
516. theory Y　Y 理论
517. thermodynamicist *n.* 热力学专家
518. thermostat *n.* 自动调温器
519. thigh *n.* 大腿，股
520. third-party involvement 第三方介入
521. thread *n.* 线，螺纹，路线 *v.* 穿过
522. threbligs *n.* 动素
523. time study board 时间研究板
524. time study form 时间研究表
525. time study 时间研究
526. TQM（Total Quality Management）全面质量管理
527. training *n.* 培训
528. Transactional Analysis（TA）人际关系心理分析
529. tray *n.* 托盘，文件盒
530. trend analysis 趋势分析
531. trial sub-stage 尝试阶段
532. two-hand chart 操作者程序图
533. unanimity *n.* 全体一致
534. unclear performance standards 绩效评价标准不清
535. underway *adj.* 进行中的
536. unemployment insurance 失业保险
537. unfair labor practice strike 不正当劳工活动罢工
538. unsafe act 不安全行为
539. unsafe conditions 不安全环境
540. unsightly *adj.* 难看的
541. upset *vt.* 扰乱
542. upstream(downstream) *n.* 上(下) 游流水线
543. validity *n.* 效度，有效性，合法性，正确性

544. value-based hiring 以价值观为基础的雇佣
545. variable compensation 可变报酬
546. VCR（Video Cassette Recorder）录像机
547. vernier caliper 游标卡尺
548. versus *prep*. 对（指诉讼，比赛等中），与……相对
549. vestibule or simulated training 新雇员培训或模拟
550. vesting *n*. 特别保护权
551. videotape *n*. 录像带
552. visualize *vt*. 想象，设想
553. voluntary bargaining 自愿谈判项目
554. voluntary pay cut 自愿减少工资方案
555. voluntary time off 自愿减少时间
556. voucher *n*. 收据，证人 *vt*. 证实……的可靠性
557. Vroom-Yetton leadership trainman 维罗姆-耶顿领导能力训练
558. wage carve 工资曲线
559. weighting result 加权结果
560. wildcard *n*. 通配符
561. WIP-Work-In-Progress 在制品
562. withdraw *v*. 撤退，收回，撤销
563. work description 工作描述
564. work factors 工作因素法
565. work in process 在制品
566. work location code 工作场所代码
567. work measurement 作业测定
568. work sampling technique 工作抽样技术
569. work sampling 工作抽样
570. work sharing 临时性工作分担
571. worker involvement 雇员参与计划
572. workers benefits 雇员福利
573. workforce *n*. 劳动力，工人总数，职工总数
574. workhead *n*. 工作台，机台
575. workholder *n*. 工件夹具
576. working up 逐步建立
577. workpiece *n*. 工件，加工件
578. wrench *n*. 扳手
579. year-end learning accomplishments 年终学习成绩

参 考 文 献

[1] Turner W C, et al. Introduction to Industrial and Systems Engineering [M]. 3rd ed. 北京：清华大学出版社, 2002.

[2] Camp L S. The Ancient Engineers [M]. Cambridge, Mass.: MIT Press, 1963.

[3] Kemper J D. The Engineer and His Profession [M]. 2nd ed. New York: Holt, Rinehart and Winston, 1975.

[4] Smith R J. Engineering as a Career [M]. 3rd ed. New York: McGrew-Hill, 1969.

[5] Emerson H P, Naehring D C E. Origins of Industrial Engineering: the Early Years of a Profession [M]. Atlanta: IE&M Press, 1988.

[6] Hammond R W. Industrial Engineering Handbook [M]. 3rd ed. New York: McGrew-Hill, 1971.

[7] Ritchey J A. Classic in Industrial Engineering [M]. Delphi, Ind.: Prairie Publishing Company, 1964.

[8] Spriegel W R, Myers C E. The Writings of the Gilbreth [M]. Homewood, Ⅲ, Richard D. Irwin, Inc., 1963.

[9] Urwick L. The Golden Book of Management [M]. London: Newman Neame Ltd., 1956.

[10] Peter Drucker. The Age of Discontinuity [M]. New York: Harper & Row, 1968.

[11] Wiener Norbert. Cybernetics, or Control and Communication in the Animal and the Machine [M]. New York: John Wiley & Sons, 1948.

[12] Shannon Claude. The Mathematical Theory of Communication [M]. Urbana: University of Illinois Press, Urbana, IL, 1948.

[13] Regh J A, Kraebber W H. Industrial Engineering and Management [M]. New York: McGraw-Hill, 1994.

[14] Schlager K J. Systems Engineering—Key to Modern Development [J]. IRE Transactions on Engineering Management, 1956, 3 (12): 64-66.

[15] McCormick, Emest J. Human Factors Engineering [M]. New York: McGraw-Hill, 1957.

[16] Christopher O'Brien. Industrial Engineering [M]. Nottingham: University of Nottingham U. K., 2003.

[17] Larry P Ritzman, Lee J Krajewski. Foundations of Operations Management [M]. 北京：中国人民大学出版社，2004.

[18] N Gregory Mankiw. Principles of Economics [M]. 6th ed. Stamford: Thomson Learning, 2011.

[19] Paul A Samuelson, William D Nord haus. Economics [M]. 19th ed. New York: McGraw Hill Higher Education, 2009.

[20] Joseph E Stiglitz. Economics [M]. 4th ed. New York: W. W. Norton, 2006.

[21] 易树平，郭伏．基础工业工程 [M]. 北京：机械工业出版社，2014.

[22] Niebel B, Freivalds A. Methods, Standards, and Work Design [M]. 11th ed. 北京：清华大学出版社，2003.

[23] Richardson W J, Pape E S. Work Sampling [M]. 2nd ed. New York: John Wiley & Sons, 1992.

[24] Hwaiyu Geng. Manufacturing Engineering Handbook [M]. New York: McGraw-Hill, 2004.

[25] Roger Timings, Steve Wilkinson. E-Manufacture: Application of Advanced Technology to Manufacturing Process [M]. New York: Pearson Prentice Hall, 2003.

[26] James A Rehg, Herry W. Kraebber. Computer-Integrated Manufacturing [M] 3rd ed. 北京：机械工业出版社，2004.

[27] Mikell P Grover. Automation Production Systems, and Computer-Integrated Manufacturing [M]. 2nd ed. 北京：清华大学出版社，2002.

[28] 威廉 J 史蒂文森. 生产与运作管理 [M]. 张群，译. 6 版. 北京：机械工业出版社，2002.

[29] Richard B Chase, Nicholas J Aquilano, F Kohert Jacobs. Prodution and Operations Management [M]. 8th ed. 北京：机械工业出版社，2002.

[30] 刘丽文. 生产与运作管理 [M]. 北京：清华大学出版社，2001.

[31] Richard B Chase, Nicholas J Aquilano, F Kohert Jacobs. Production and Operations Management: Manufacturing and Service [M]. 北京：机械工业出版社，2000.

[32] James A Fitzsimmons, Mona J Fitzsimmons. Service Management: Operations, Strategy and Information Technology [M]. 北京：机械工业出版社，2002.

[33] 叶春明. 生产计划与控制 [M]. 北京：高等教育出版社，2005.

[34] Donald J Bowersox, David J Closs, M Bixby Cooper. Supply Chain Logistics Management [M]. 北京：机械工业出版社，2003.

[35] Fred E, Meyers Matthew P, Stephens Meyers. Manufacturing Facilities Design and Materials Handling [M]. 2nd ed. 北京：清华大学出版社，2002.

[36] 宋伟刚. 物流工程及其应用 [M]. 北京：机械工业出版社，2003.

[37] Matthew G Anderson. Strategic Sourcing [J]. International Journal of Logistics. January 1998: 1-13.

[38] [作者不详] Grocery Distribution. Loading Dock 2000. October 1998: 28-31

[39] [作者不详] The Importance of Proper Sequence. Warehousing Forum. June 1997: 1-2

[40] Sanders M S, McCormick E J. Human Factors in Engineering and Design [M]. 7th ed. 北京：清华大学出版社，2002.

[41] Charlotte E N. Analysis and Evaluation of Working Posture [J]. Ergonomics of Workstation Design, 1983: 1-18.

[42] Stephen P. Anthropometry, Ergonomics and the Design of Work [M]. 2nd ed. London: Taylor & Francis. 1998.

[43] Howard Gitlow. Quality Management [M]. 3rd ed. 北京：清华大学出版社，2004.

[44] Juran J M, Gryna F M. Juran's Quality Control Handbook [M]. 4th ed. New York: McGraw-Hill, 1988.

[45] Anderson. David L. Management Information Systems [M]. 北京：清华大学出版社，2001.

[46] Robert A Schultheis, Mary Summer. Management Information Systems [M]. 北京：机械工业出版社，1998.

[47] 王恩波，等. 管理信息系统实用教程 [M]. 北京：电子工业出版社，2002.

[48] Lin Grensing-Pophal. Human Resource Essentials: Your Guide to Starting and Running the HR Function [M]. 2nd ed. New York: SHRM/Davies-Black Publishing, 2002.

[49] Ian Price. Human Resource Management in a Business Context [M]. 2nd ed. Singapore: Thomson Learning, 2004.

[50] 樊玉敬. 科技文献检索与利用 [M]. 北京：煤炭工业出版社，2002.

[51] Hart C. Doing a Literature Search: a Comprehensive Guide for the Social Sciences. London SAGE Publication Ltd., 2001.

[52] Pritchard E, Scott P. Literature Searching in Science, Technology, and Agriculture [M]. Westport, Conn. GREENWOOD PRESS Westport, 1996.

[53] Gash S. Effective Literature Searching for Research [M]. 2nd ed. England: Gower., 2000.

[54] Lee Pao M. Concepts of Information Retrieval [M]. Englewood: Libraries Unlimited, Inc. Englewood, Colorado, 1989.

[55] Literature Searching: a User Guide, October 2002. www. csp. org. uk.

[56] 严俊仁. 科技英语翻译技巧 [M]. 北京: 国防工业出版社, 2000.

[57] 范武邱. 实用科技英语翻译讲评 [M]. 北京: 外文出版社, 2001.

[58] 秦获辉. 实用科技英语写作技巧 [M]. 上海: 上海外语教育出版让, 2001.

[59] 王建武, 李民权, 曾小珊. 科技英语写作——写作技巧·范文 [M]. 西安: 西北工业大学出版社, 2000.

[60] 丁往道, 吴冰. 英语写作手册 [M]. 修订本. 北京: 外语教学与研究出版社, 2000.

[61] James R Holt. Comparing Performance of Different Production Management Approaches (Downloadable from website http: //public. wsu. edu/ ~ engrmgmt/holt/em530/AP)

[62] Luis Vildosola Reyes. Circular Variable Work in Process 2002. Macroeconomics, EconWPA 0203004, EconWPA. (Downloadable from website https: //ideas. repec. org/p/wpa/wuwpma/0203004. html)

《工业工程专业英语》第2版（周跃进　任秉银主编）

信息反馈表

尊敬的老师：

您好！感谢您多年来对机械工业出版社的支持和厚爱！为了进一步提高我社教材的出版质量，更好地为我国高等教育发展服务，欢迎您对我社的教材多提宝贵意见和建议。另外，如果您在教学中选用了《工业工程专业英语》第2版一书，我们将为您免费提供与本书配套的PPT课件。

一、基本信息

姓名：________　性别：________　职称：________　职务：________

邮编：________　地址：____

任教课程：____________　电话：______—________（H）________（O）

电子邮件：____________________　手机：________________

二、您对本书的意见和建议

（欢迎您指出本书的疏误之处）

三、您对我们的其他意见和建议

请与我们联系：

100037　机械工业出版社·高等教育分社·管理与经济编辑室　裴编辑　收

Tel：010—88379539

Fax：010—68997455

E-mail：cmppy@163.com